FOLLOWING ST. FRANCIS

FOLLOWING ST. FRANCIS

John Paul II's
Call for Ecological Action

<small>WRITTEN AND COMPILED BY</small>

Marybeth Lorbiecki

<small>FOREWORD BY</small>

Bill McKibben

Rizzoli
ex libris

Published in the United States of America in 2014
By Rizzoli Ex Libris, an imprint of
Rizzoli International Publications, Inc.
300 Park Avenue South
New York, NY 10010
www.rizzoliusa.com

2014 2015 2016 2017 / 10 9 8 7 6 5 4 3 2 1

Distributed in the U.S. trade by Random House, New York
Printed in U.S.A.

ISBN-13: 978-0-8478-4271-1
Library of Congress Catalog Control Number: 2013943441

*An alliance between all those who want goodness
is extremely urgent today.
Humanity and the world are at stake
and they are endangered as never before.
As I said before:
Protect the world,
the beautiful, endangered world.*

POPE JOHN PAUL II, ADDRESS TO REPRESENTATIVES OF SCIENCE,
ART, AND JOURNALISM, SALZBURG, AUSTRIA, JUNE 26, 1988

Contents

❧ Foreword: Reorientation and Action

BILL MCKIBBEN

This is the profoundly right moment for this book. It comes as all of us—even Methodists like myself—are feeling the excitement of a new pope, Francis, who seems to have broken the mold, and to speak with a freshness and a candor that the world is eager to hear. One can't help but be charmed and challenged by his actions, from washing the feet of Muslims to refusing to judge our gay and lesbian brothers and sisters. And one senses in his choice of a name that this new approach will extend to our relationship not just with each other but with the planet God has given us.

But few things spring up absolutely afresh in human history—and it's doubly rare in the Church! So it's very good to be reminded in this book that the current excitement has precedent both fairly recent and fairly ancient. Marybeth Lorbiecki does a lovely job of tracing Pope John Paul II's ecological insights, which were acute and deep—the product not only of his boyhood in the natural world and his faith, but also, perhaps, of the chance to watch how the insane materialism of his communist homeland had destroyed so much of that world. Right from the start of his papacy, he understood the core of the problem and its urgency: as he said in 1979, each of us needs "a deep personal conversion in his or her relationship with others and with nature."

The ancient source of that understanding harkens back to the first books of the Bible and onward, then on through St. Francis of Assisi, perhaps the most beloved of the line of saints, and certainly one of the most unlikely. His surpassing ability to see God in the world around him (human, beast, and even rock), adds enormous dimension to the Christian story; the radicalism of his message has in some sense lain buried until it was needed in this moment of ecological crisis.

The scriptural and moral principles the pope laid out have points of universal resonance, showing how spirituality and ecology intersect—and how the many different ecological issues are related. They come after a century in which ecological insight really flowered; now the deeper meanings of that new science are being explored, and powerfully so in the case of John Paul II.

Simply because this message is needed, of course, doesn't mean it will be acted on. Pope John Paul II did say many powerful things; it must be noted, however, that he didn't turn the Church into an effective fighting force for ecological progress. It was left to Patriarch Bartholomew of the Orthodox Church to show, one watershed at a time, how religious leaders can rally society to action out of the wellsprings of their faith.

One hopes—hopes against hope, but then what is the Holy Spirit for?—that the current pope will truly channel the beautiful radicalism (or is it conservatism?) of his namesake, and that he will build on this faith legacy of the newly named St. John Paul, and on the powerful science that the exercise of our civilization's reason has produced. If so, then the words in this book will prove powerful preludes to some of the most needed deeds in the short history of the human species.

ᴗᴗ Author's Note:
Why and How to Read this Book

In his book *Blessed Unrest*, entrepreneur and author Paul Hawken writes:

> It has been said that we cannot save our planet unless humankind
> undergoes a widespread spiritual and religious awakening. In other
> words, fixes won't fix until we fix our souls as well. So let's ask our-
> selves this question: Would we recognize a worldwide spiritual awak-
> ening if we saw one?... What if there is already in place a large-scale
> spiritual awakening and we are simply not recognizing it?

His book suggests we *are* in the midst of one, and we need only notice
and join in.

Pope John Paul II, newly named a Catholic saint, was a follower of St.
Francis, often quoting him or referring to him as a model for us all. In his
own time, St. Francis of Assisi was a spiritual movement unto himself: a man
radically and wholeheartedly devoted to Christ, creation, and the poor. His
power surpassed that of kings and popes who are no longer much remembered, and his work lives on, expanding to inspire new generations and contemporary popes—especially Pope John Paul II and Pope Francis. His legacy
has never been more relevant.

John Paul II's Spiritual and Ecological
Primer of Principles and Actions

Of all the dozens of books published on Pope John Paul II, not one has
focused on his persistent call to radically change our orientation toward the
earth in the name of our Creator and to take action to protect the systems
and species upon which life depends—in other words, to follow St. Francis.
This book is the only synthesis of John Paul II's writings, speeches, and
encyclicals on the subject. In the decade I spent compiling this collection, I
was struck again and again by the late pope's deep insights into the human

psyche and the complex ecological interrelationships that exist between the physical, economic, emotional, social, moral, and spiritual realms. As the New Testament letter to the Corinthians notes, "There are many parts, but all one body."

John Paul II was asking us to rethink, through the window of ethics and faith—specifically through Judeo-Christian Scripture and Catholic teachings—who we are in this body of the earth, and he gives us many thought-provoking ways to do so. His counsel, however, is certainly not just for Catholics, nor even just for Christians. Around the world, religious and secular communities of all kinds are articulating similar spiritual calls to care for our natural systems, habitats, and species. However, not all are explaining the structural interconnections.

Long before others did, and to the surprise of many, John Paul II described the interactions between poverty and ecological degradation; between war, environmental destruction, and long-term loss of land fertility; between deforestation, drought, and desertification; between consumerism, waste, economic recession, personal depression, and habitat loss. He spoke against climate change denial and in favor of switching to sources of energy that work with the earth's forces instead of against them. He advocated limiting population through responsible parenthood, educating girls, and engaging women in all levels of leadership for more community and environmental decision making. He called for protecting indigenous people's rights on their lands—and looking to them for ecological wisdom.

His vision was based on a belief in the power of human beings to change and do better, alone and together, through the grace of God and the guidance of science and ethics. He highlighted the opportunities present in our environmental crises for multilevel transformations and ways to sow the seeds of a new springtime of peace with God, and peace with all of God's creation. His insights can be considered and applied from whatever faith or spiritual perspective a person lives by.

This book unfolds as follows:

> *Part I: In the Footsteps of Francis* introduces the life and example of St. Francis, and the forming of John Paul II's ecological insights.

Part II: The Ecological Emergency and Moral Crisis: The Scriptural Call for Urgent Action gives the ethical, scriptural, Catholic, social, and scientific rationale for Pope John Paul II's beliefs.

Part III: Ecological Violence: The Key Issues offers Pope John Paul II's specific counsel on the twelve major contemporary ecological issues.

Part IV: Ecological Conversion and Restoration provides steps for our culture to move forward, three significant examples of restorative hope from St. John Paul, and a homily on care of creation given by Pope Francis.

Part V: Resources and Speeches contains a sampling of organizations to turn to for action and prayer, a list of key papal works addressing ecology, and four ecological speeches by Pope John Paul II.

This book is, of course, not a summation of all that Pope John Paul II taught, believed, or held dear, or of the basic tenets of Catholicism. Much has been written on all these topics already, and they are assumed in this book. This book is also by no means comprehensive even on his teachings on ecology—the assembly of quotations, however, does present the major themes and recommended actions for caring for God's creation that flowed through his encyclicals, speeches, sermons, and addresses to the United Nations, to youth, and to groups of all kinds in countries around the world. It also highlights the relevant teachings of others, especially Pope Francis. However, unless otherwise noted, all quotations that follow are by John Paul II, and any italics used for emphasis are the pope's own. Also, the noninclusive practice of using the word "man" instead of person, and "men" instead of people when both sexes are intended is part of Pope John Paul II's traditional, formal language and therefore has not been edited.

Some readers may be concerned that some quotations are highlighted outside of their context. It couldn't be any other way. Though John Paul made a number of key ecological addresses, many of his teachings on God's creation are woven like sunlit threads through his other teachings on Catholicism, life, and family issues. To present his ecological teachings fully, as well as how they fit together, they had to be collected and organized.

Since he developed each ecological theme in far greater depth than can be presented here, at the end of this book there is a list of significant documents and a few key addresses so you can hear his voice in context.

The Prophetic Power

His voice is one that is sometimes lyrical and profound, sometimes practical, and sometimes prophetic. In view of contemporary politics, it's hard to believe that he was defining the ecological crisis as a faith and moral crisis as far back as the 1980s. Pope John Paul II's recommendations are not spiritual lozenges to comfort and soothe, but radical concepts—precise moral counsel that scissors through any cloak of "rights" used to cover irresponsibility or selfishness, whether worn by nations, corporations, religious organizations, or individuals. The pope's guidance aims to move us toward building the interfaith, intersecular, international solidarity necessary to wrestle with challenges as intimidating as global warming and habitat loss, poverty and war—developing not one answer but a kaleidoscope of diverse, colorful solutions working locally, nationally, and internationally. And in the process, proving ourselves to be wiser, truer, kinder, more inventive—and our lives fuller and more meaningful—than we ever believed possible.

Like those of St. Francis, the words of St. John Paul II are decorated by hope and joy—in God, nature, and the human spirit. He believes in us far more than we often believe in ourselves. These words give wind to the sails of many groups and important movements working today, including the Franciscan Action Network, the Catholic Climate Covenant, the Charter of Compassion, the Earth Charter, the UN Millennium Development Goals, the Greenbelt Movement, the numerous environmental and social justice nonprofits, and 350.org, the grassroots global network for united action to slow climate chaos.

Humble and Blessed

In his lifetime, St. John Paul II was acknowledged around the world for his mysticism, warmth, and personal holiness. Yet, like all humans, he made missteps, such as how he dealt with the problem of sexual abuse by clergy, and had his blind spots. Some may accuse him as pope of not adequately practicing what is outlined in this book; or not applying these ecological precepts to the structures of the Roman Catholic Church; or not rallying believers to take sufficient action on significant problems. This book is not meant to focus on the life of John Paul II or on his specific papal decisions, but instead to pass on the extensive ecological and spiritual wisdom he left us as a gift and faith mandate.

A humble man in life, the pope would have been the first to admit that he was not perfect, as he admitted that the Church itself was not perfect. He publicly apologized ninety-four times for historic wrongs done by the Roman Catholic Church and tried to address some of the contemporary ones. He reminded us all that we need humility and repentance in the face of the Divine, and that we do not control God's Spirit on earth or within the vast, complex mystery of this universe, of which we are proportionally less than an atom. Yet what we *can* control is who we want to be—as individuals and as communities of faith. Do we want to be co-creators with God to save this "beautiful, endangered world" or not?

My Own Franciscan Journey

As a child, I attended schools where the nuns (Benedictine, Franciscan, and Sisters of St. Joseph of Carondelet) taught that God loves all of nature and all of us—especially the poor of this earth—no matter our color or religion, and that each of us is called to imitate Christ and care for all. We celebrated Earth Day, and as a middle schooler, I even wrote a letter to President Nixon asking him to take care of the environment. My parents reinforced these values by having us spend time as a family canoeing, hiking, and camping—always teaching us to leave a campsite (and this world) cleaner and better than we found it.

However, as I grew older, I saw less and less of this environmental consciousness in Catholic parishes, schools, and media. Although the U.S. Conference of Catholic Bishops issued strong statements on the environment, and Catholic social teachings included care for nature as part of stewardship, there was little to no diocesan or parish preaching, teaching, or action to put these principles into practice. Pro-life activists did not include combating ecological threats to the unborn or to future generations. Outdoor masses became fewer and fewer. To my shocked dismay, I met many Catholics— including clergy, bishops, and pro-life advocates—and other Christians who were *vehemently anti*-environment. They called environmentalists "pantheists" who "adore" nature, valuing nature *over* humans and over unborn children! Yet what kind of world would we be offering these newborn babies as the climate is shifting to make the planet increasingly inhospitable? And what of all the children being miscarried or born with birth defects from chemical pollutants?

Looking deeper, I wondered what had happened to Catholicism and Christianity I had known. How had partisan politics been allowed to create false splits between life issues?

A Spiritual Journey of Hope

Challenged, I turned to the Bible for counsel and there discovered a definite and insistent faith mandate threading throughout Scripture, calling for us to tend well God's lands, waters, animals, and plants. Not a lot of gray area. Just consider the speech after Noah's ark in Genesis 8 and 9, when God puts humans in the same covenant with the animals and all living creatures—saying it five times lest we not pay attention. Leviticus 26 spells out the benefits if we remember God's land covenant and precepts: "I will give you rain in due season, so that the land will bear its crops, and the trees their fruit...I will set my Dwelling among you, and will not disdain you. Ever present in your midst, I will be your God, and you will be my people..."

However, if we forget and break the covenant: "I, in turn, will give you your deserts. I will punish you with terrible woes—with wasting and fevers to sap the life....I will make the sky above you as hard as iron, and your soil as hard as bronze, so that your strength will be spent in vain; your land will bear no crops, and its trees no fruit." There will be famine, pestilence, and wars, with cities laid to ruin. "Then shall the land retrieve its lost sabbaths..."

To my ears, the warnings sounded frighteningly like the predictions coming from the scientists and the military about the consequences of climate change. I realized that God had built consequences for bad behavior right into this planetary household's own structural systems. You follow the system, you thrive. You work against it, and everything falls apart around you a little at a time. And this angers God. Just read from the prophet Jeremiah and others. The more we know about ecology and its connections to social systems, the more Leviticus 26 sounds like the natural and social consequence of willful ecological destruction.

The repeating underlying message in Scripture was: value as I, the Lord God, value (*and God saw that it was good*); and do your best to care for the garden of creation as I, the Lord your God, have directed you; care for the poor and vulnerable as I have cared for you; love others as I have loved you; love your neighbor (and strangers) as yourself; love your enemies and pray for them. Remember your God every day in all things, and spend a full day of rest once a week to remember and honor your God (this is repeated often). Give your lands and animals time to rest, too. Set aside portions of land as sanctuaries and wilderness for God's uses and those of God's animals and plants. Serve others. Pause often each day to remember that all flows from God and returns to God. Pray outdoors as the prophets and Christ did; praise God in thanks, awe, humility and wonder. And when you don't do these things, or you make mistakes, then repent, reconcile, and change your ways. Serve others. Feed others. Heal others. Renew the face of the earth.

This was, obviously, all good counsel. But I wasn't seeing much of the earth-care teachings among believers in a Creator—Christians particularly *as* Christians, Jews *as* Jews, or Muslims *as* Muslims, etc. There were many religious people at work on ecological issues in secular organizations but not, it seemed, specifically or publicly because of their faith. I did not see a lot of groups from churches, temples, and mosques getting to work together on climate issues, reforestation, lake, river, ocean issues, and the like.

Delving into the science, I examined the life and literary writings of the author, ecologist and environmental ethicist Aldo Leopold and wrote a biography of him. Like many of the U.S. founding fathers, he seemed to have a kind of Deist belief in a Creator who started the natural processes of the planet and then left them to their own devices and to human ethics. Leopold was disillusioned by organized religion's ignoring of ethics when it came to the land and other species.

Then I stumbled onto the teachings of Pope John Paul II on ecology highlighting St. Francis as a model, and I was thrilled. Some of John

Paul's teachings closely resembled Leopold's. The pope's counsel was fresh and surprising—firmly based in Scripture *and* science, heart-felt, religiously strident, deeply spiritual and challenging, beautiful, inspiring, and hopeful. Why wasn't anyone talking about it? Why hadn't anyone listened?

Now that he has been recognized as a saint, it may be time. We may be ready to listen and be part of the spiritual awakening and action he was calling forth. And the rest of the earth is certainly ready. For this book is not really about St. John Paul II, or Pope John Paul II, or about the Catholic Church. It is about us and our choices.

It is about the future of the planet and life itself.

PART I

IN THE FOOTSTEPS OF FRANCIS

*Everyone remembers the stories of St. Francis
joyfully speaking with the animals,
urging them to respect others and to praise the Creator.
This example is particularly urgent for our times
when, without the slightest concern,
man is slowly destroying the environment
that the Creator had prepared for him.*

POST-ANGELUS: NATIONAL DAY OF ECOLOGY AND ZOOLOGY, MARCH 28, 1982

Look to the future with hope,
and set out with renewed vigor
to make this new millennium
a time of solidarity and peace,
of love for life
and respect for God's creation.

PILGRIMAGE TO MALTA, MAY 8, 2001

The Legacy of St. Francis and the Ecological Formation of St. John Paul II

Shortly after the 2013 election of Pope Francis, when the press asked him why he chose his name, he said he was thinking about the lives of the poor and the horrors of war, and St. Francis of Assisi came to mind.

Who Was St. Francis?

As a young man in the early 1200s, Francis was a rich kid in Assisi, enjoying a life of wine, women, and song thanks to his father's money. After a brief stint as a soldier, and far longer periods as a prisoner of war and an invalid, he started to have visions. He rambled more and more often through the woods in prayer. Money and adventure were no longer enough, especially after the sight of a poor leprous man sank a dart of compassion into Francis's heart. Praying for direction in the small stone church of San Damiano, he heard Jesus on the cross say to him three times, "Francis, go and rebuild my house, which you see is falling down."

The young man took Christ seriously and literally. After a conflict over money with his businessman father, Francis stripped down naked in Assisi's public square and tossed his fancy clothes and money back at him, renouncing him before the bishop and his neighbors. (Francis was nothing if not dramatic!) From then on, barefoot and dressed in a ragged cloak, he rebuilt the little church stone by stone, then wandered through the countryside fixing other churches and preaching repentance and the love of God, sharing what little was given to him with the poor. He was never a priest, nor did he aspire to be one. Yet he was so passionate and inspiring, people followed him. Not originally intending to, he launched three Catholic orders of serving communities (one of friars and monks, one of sisters with his dear friend Clare, and one of lay people). These have spawned many more. He lived mostly outdoors, sleeping indoors only in illness or out of kindness to a host. He radiated joy, song, laughter, peace, love, compassion, and generosity to all he met, including the birds, animals, and plants. A poet and troubadour, he wrote the famous "Canticle of the Creatures," singing God's praises outdoors with these brothers and sisters.

The nativity set we all know, with the animals at the manger, was also his work. It was 1223 and as a visiting preacher, he wanted to make Christ's birth at midnight Mass more real for everyone, so he arranged with a farmer to bring a manger, hay, oxen, and a donkey to a hillside cave opening. A woman playing Mary held a wax baby while a man as Joseph bent over them in care. Real shepherds gathered at the edges with their sheep to attend Mass. Francis preached so movingly that for a few moments, it seemed that the baby transfigured into the real Jesus. (The straw was later said to have attained the miraculous qualities of curing the local cattle of disease.) So close was Francis to Christ that the next year, when he was praying before the cross asking to share in the Lord's suffering, he received Jesus's crucifixion wounds on his own body, the first recorded occurrence of the stigmata.

It is Catholic tradition to have the blessing of the animals on St. Francis's feast day (October 4), as he was named their patron saint. Francis is one of Christianity's best-loved saints, respected even among many non-Christian cultures. This was the case even during his lifetime. During the Fifth Crusade, in the midst of backstage negotiations to stop the fierce fighting between the Christians and Muslims, he bravely walked across the battle lines to speak to the sultan of Egypt, Al-Kamil, hoping to convert him. No conversions took place then, but a generous dialogue between the sultan and the prisoner did, and the two men ended in mutual respect, becoming friends of different religions. (Later, only Franciscans could stay in Egypt and the Holy

Lands after the fall of the Crusader empire. Some Catholic traditions say that years after their meeting, when Al-Kamil felt he was dying, he converted to Christianity on the promise of Francis's prayers and intercession.)

The New Pope's Decision

So why did the new pope specifically take the name of St. Francis of Assisi? "For me," Pope Francis told the crowds, "he is the man of poverty, the man of peace, the man who loves and protects creation; these days we do not have a very good relationship with creation, do we?"

The overwhelming answer is no, we don't. When thousands of sea lion pups wash up on the coast of California starving for lack of fish, the answer is no. When deserts are multiplying as fast as polar ice sections are disappearing, and species are going extinct at the rate of 150 to 200 a day (a thousand times the natural rate), the answer is no. When the poor are forced to pay for clean water to drink because there is none in their faucet or river, the answer is ashamedly no. When the economies of nations are unbalanced by extreme profits for corporate officers and stockholders while everyday workers are suffering layoffs, poor wages, unsafe working conditions, and toxic chemicals in their air, water, and food, the answer is no. And when citizens and soldiers are suffering through wars over oil and other natural resources, the answer can only be NO.

In these times, hope sometimes seems in scarce supply. Yet despite these sad developments, there has emerged an unexpected delight in Pope Francis because he chooses to live simply, to engage in interfaith dialogue, and to work to reform the Church and the world, calling all back to their roots in God and the earth. He tweets, runs from his security guards to embrace those with special needs, and washes the feet of Muslim girls. Sounds like an updated Francis. He says the Church is too self-referential and must be more humble, less sure of itself, and more sure of the Mystery of God. Francis again. And just like St. Francis, Pope Francis reminds us that God the Creator loves us all, no matter our beliefs. In a now-famous homily, Pope Francis told a story of a Catholic asking a priest about the salvation of atheists:

> The Lord has redeemed all of us, all of us, with the Blood of Christ: all of us, not just Catholics. Everyone!

Father, the atheists?

Even the atheists. Everyone! And this Blood makes us children of God of the first class. We are created children in the likeness of God and the Blood of Christ has redeemed us all.

And we all have a duty to do good. And this commandment for everyone to do good, I think, is a beautiful path towards peace. If we, each doing our own part, if we do good to others, if we meet there, doing good, and we go slowly, gently, little by little, we will make that culture of encounter: We need that so much. We must meet one another doing good.

But I don't believe, Father, I am an atheist!

But do good: We will meet one another there.

CHAPEL OF THE DOMUS SANTA MARTA, VATICAN, MAY 22, 2013

In this spirit of St. Francis, we might be ready to band together as St. John Paul II and Pope Francis advised, beyond religion and nation, to face our ecological crises with action, to meet each other in doing good.

The Ecological Crisis Is a Moral Crisis

In his first World Day of Peace Message on January 1, 1979, Pope John Paul II noted that if the world wanted peace, we'd have to care for the environment on which we all depend. In the next twelve years, he highlighted this same theme alongside other pressing world issues in three encyclicals, one exhortation to lay people, two addresses to the United Nations, one message to the European Parliament and six to sessions of the Pontifical Academy of Sciences, one apostolic letter to youth, and at least ten other key messages, addresses, and sermons in English to various audiences (and who knows how many address in other languages!). It recurred especially in his many messages to young people.

By the time Pope John Paul II delivered the World Day of Peace speech in 1990, "Peace with God, Peace with All of God's Creation," he was clearly frustrated that there was no united outrage and change of direction in his flock or the world. So twice in the same speech he admonished:

"I wish to repeat: The ecological crisis is a moral crisis!" (the emphasis is the pope's own).

This was an incredibly bold statement for any world leader, much less a pope, to make. But he chose his words thoughtfully, using the truth to shock people out of complacency. And he continued to use the statement until his death. Pope John Paul II was disturbed that the world and church leaders were not understanding how urgent these ecological crises were and how responsible they, and we, were for doing something. We are facing ongoing worldwide sin, he taught, both personal and communal, and we have to quit making rationalizations and excuses, then repent and get to work. He warned that whether we acknowledge it or not, we have the power to destroy life as we know it, to bring about *"the danger of an environmental holocaust"* (ADDRESS TO YOUNG PEOPLE IN THE DOLOMITES, JUNE 28, 1991).

Pope John Paul II explained that caring for the environment is not a matter of politics or personal interest, but the essential duty of every human being, government entity, religious group, and humanity as a whole. He made this very clear in his message to the Regional Council of Lazio, the region in which Rome is located:

> As you know in the recent message for the World Day of Peace, I called to the attention of every person of goodwill a serious issue— the problem of ecology—recalling that in finding a solution, we must direct the efforts and mobilize the will of citizens. An issue like this cannot be neglected—for it is vital for human survival—nor can it be reduced to a merely political problem or issue. It has, in fact, a moral dimension which touches everyone and, thus, no one can be indifferent to it.
>
> …humanity is called to establish a new relationship of attentiveness and respect towards the environment. Humanity must protect its delicate balances, keeping in mind the extraordinary possibilities but, also, the formidable threats inherent in certain forms of experimentation, scientific research and industrial activity—and that must be done if humanity does not want to threaten its very development or draw from it unimaginable consequences.... Ecological problems enter into

everyone's home, they are discussed in the family circle and people wonder what tomorrow will be like.

We must, therefore, mobilize every effort so that each person assumes his or her own responsibility and creates the basis for a lifestyle of solidarity and brotherhood.

ADDRESS TO REGIONAL COUNCIL OF LAZIO, FEBRUARY 5, 1990

The Call for Ecological Conversion

The solution? We all need "an ecological conversion."

On the basis of the covenant with the Creator…each one is invited to a deep personal conversion in his or her relationship with others and with nature.

ADDRESS TO SEMINAR "SCIENCE FOR SURVIVAL AND SUSTAINABLE DEVELOPMENT,"
PONTIFICAL ACADEMY OF SCIENCES, MARCH 12, 1999

He explained that each human being, by his or her birthright, has an inherent "ecological vocation," just as the earth has an inherent structure of ecological relationships and systems of which we are a part.

When the ecological crisis is set within the broader context of the search for peace within society, we can understand better the importance of giving attention to what the earth and its atmosphere are telling us: namely, that there is an order in the universe which must be respected, and that the human person, endowed with the capability of choosing freely, has a grave responsibility to preserve this order for the well-being of future generations.

1990 WORLD DAY OF PEACE MESSAGE,
"PEACE WITH GOD THE CREATOR, PEACE WITH ALL CREATION"

But how did the garden of John Paul II's theology grow?

The Roots of John Paul II's Theology of Ecology

Long before becoming a pope, Karol Józef Wojtyła was a man in love with the outdoors—a hiker, camper, kayaker, swimmer, skier, and soccer player (as well as an actor and poet). After earning his doctorate and serving as a parish priest, he began teaching and lecturing on love, responsibility, and ethics at the Catholic University of Lublin. During short respites, he led groups of students, both male and female, out into the mountains for retreats and leisure. Affectionately, the students dubbed the noncollared priest "Uncle" for fear Communist spies might hear him called "Father" (the Communists had banned priests from leading youth groups). On these trips, students learned to live lightly and simply in communion with nature, with time for solitude and prayer, and long evenings at the campfire surrounded by singing, stories, and laughter.

By the birds' first canticles at dawn, Uncle would be out paddling, stopping at particular points to read and contemplate Scripture, then he'd return to say Mass for the group to start their day. Always creative, he bound two kayak paddles together for a cross and laid out his altar on overturned kayaks. It was during a camping trip in the lakes region of Poland in 1958 that he got word to return, for he was to become a bishop, at the age of thirty-eight. Onlookers claim that when the bishop's miter was laid on his head, the sun burst through a cloudy veil, blazing the stained glass and splattering him with rainbows of color.

In 1978, John Paul II spent his first year as pope preaching sermons on the creation stories of Genesis and on the God-given relationships and responsibilities they revealed (he returned to these passages in 1986, as ecological issues were getting worse). He based his teachings not just on the Old Testament stories and prophets, but on Christ, the epistles, and contemporary Catholic teachings from Pope John XXIII and the Second Vatican Council onward through each successive pope. Pope Paul VI had advocated to the UN for good stewardship of the environment in 1970 (with assistance from his right-hand man, Cardinal Wojtyła). Even Pope John Paul I, with so few days in office, had warned:

> The danger for modern man is that he would reduce the earth to a desert, the person to an automaton, and brotherly love to a planned collectivization, often introducing death where God wishes life.
>
> FIRST MESSAGE TO COLLEGE OF CARDINALS AND TO WORLD,
> AT CONCLUSION OF A MASS, SISTINE CHAPEL, AUGUST 27, 1978

For life models, John Paul II looked to the numerous saints who loved and contemplated nature for clearer understandings of the Creator: Clare and Benedict, Irenaeus and Augustine, and Thérèse of Lisieux—as well as popes who became saints, such as Clement I and Leo the Great. However, the saint to top them all for care of creation was, of course, St. Francis.

In 1979, as one of his first major acts as pope, John Paul II proclaimed Francis the patron saint of those who promote ecology: "For he, in a special way, deeply sensed the universal works of the Creator and, filled with a certain divine spirit, sang that very beautiful 'Canticle of the Creatures'" and offered "an example of genuine and deep respect for the integrity of creation" (APOSTOLIC LETTER, *INTER SANCTOS*, NOVEMBER 29, 1979; MESSAGE FOR WORLD DAY OF PEACE, 1990).

Francis respected even the insects and plants, the rocks and hills, the sun and the moon, feeling that they, too, in their own ways, were praising God with their lives and existence. Francis preached that offering respect, dignity, compassion, and love to nonhuman forms of life did not diminish humans but enlarged them and made them more like Christ, and more truly themselves, in harmony with the Creator's love. The pope affirmed this:

> When man habitually loves and respects lower creatures, he will also learn to be more human with his equals. I am therefore happy to encourage and bless those who work to assure that, in the Franciscan spirit, animals, plants, and minerals be considered and treated as "brothers and sisters."
>
> POST-ANGELUS GREETING TO PARTICIPANTS IN
> TERRA MATER [MOTHER EARTH] SEMINAR, OCTOBER 3, 1982

Since Francis was already an international icon, crossing religious and cultural barriers, John Paul II meditated on the relevance of his example for all:

The witness of Francis leads those of today not to plunder nature, but to assume responsibility for it, taking care that it all remains healthy and unharmed, ... to offer future generations an environment that welcomes their presence and in which they can be at ease.

POST-ANGELUS: NATIONAL DAY OF ECOLOGY AND ZOOLOGY, MARCH 28, 1982

On Faith and Science: Two Sides of Seeking

Pope John Paul II did not rely only on Scripture and saintly models, but on science. He called many conferences of the Pontifical Academy of Sciences to study and report their consensus on varied ecological issues, such as geosphere and biosphere interactions and climate change, population, tropical forests and conservation of biodiversity, sustainability, alternative energy, chemical pollution, and other key topics. The Pontifical Academy of Sciences, like the U.S. National Academy of Sciences, comprises renowned experts in their fields, with many Nobel Prize winners among its ranks. For John Paul II, and the contemporary Catholic Church, scientific inquiry is just another tool to seek out knowledge of the Mystery of God.

Understanding the different but complementary roles of science and faith was so important to Pope John Paul II that he wrote a whole encyclical on it: "On Faith and Reason." Just one year after becoming pope, he addressed an audience of professors and students at Cologne Cathedral in honor of the religious scientist Albert the Great (Albertus Magnus), to express regret over some of the Church's past antagonism and fear toward science. He used this as an invitation to open the dialogue more fully between science and faith, mourning the fact that many people

> still feel the weight of those notorious conflicts that arose from the interference of religious authorities in the process of the development of scientific knowledge. The Church remembers this with regret, for today we realize the errors and shortcomings of these ways of proceeding. We can say today that they have been overcome....
>
> NOVEMBER 15, 1980

Evolution and Science: Truth Cannot Contradict Truth

The pope proved how seriously he respected scientific inquiry when he commissioned the Pontifical Academy of Sciences to look into evolution. In October 1996, he thanked the study group and admitted that at "a first sight there are apparent contradictions" between revelation in Holy Scripture and the discoveries of science, but on deeper examination, there are not, for "truth cannot contradict truth." He explained that science discovers what can be physically observed of this natural world, detailing what nature does over time and predicting what is to come, but it cannot delve into the deeper spiritual questions of why and how. Nor can science offer the meaning or source of nonmeasurables such as love, prayer, sensing of the sacred, and so on. This is the realm of revelation, Scripture, theology, philosophy, and psychology.

Once we take into consideration the different goals and methods used between science and sources of inspiration, he assured the world, it "makes it possible to reconcile two points of view that would seem irreconcilable." Because so much observational evidence has come to light that substantiates evolutionary theory, John Paul II concluded: "The convergence, neither sought nor fabricated, of the results of work that was conducted independently is in itself a significant argument in favor of this theory."

He did, however, distinguish that the soul formed separately without evolution: "If the human body has its origin in living material which pre-exists it, the spiritual soul is immediately created by God" (MESSAGE TO PONTIFICAL ACADEMY OF SCIENCES, OCTOBER 22, 1996).

His pronouncement on evolution rocked the world. (Some even accused him of being the anti-Christ.) Yet for the pope, it wasn't earth-shattering but simply a matter of the accumulation of evidence. While some Catholic and Christian leaders were bashing the biological and ecological scientists, Pope John Paul II was consulting them and thanking environmental groups, such as the World Wildlife Fund, for their work on behalf of God's creation.

It is not a coincidence that science writer Bill McKibben, who is himself a faith-filled Methodist, had released the now-classic book *The End of Nature* in 1989 just before John Paul II's 1990 World Day of Peace speech in which he warned about our power to destroy the earth and the moral crisis that entailed. In *The End of Nature*, McKibben made an impassioned plea for people to radically reorient themselves in their view of nature and their use of it, for we were eroding the building blocks of the earth's structure. His, too, was a moral argument, but it primarily used scientific findings and universal ethics to support it, rather than faith. Since then, McKibben has turned, like the pope, to people of Christian, Jewish, and Muslim faiths and others to more fully involve them in the struggle to slow climate change.

With some of his graduate students, McKibben organized a worldwide, nonpartisan, grassroots movement for ecological action against climate change called 350.org. This group has successfully engaged people throughout the world from varied cultures, religions, nations, civic groups, nonprofits, and governments to help protect the climatic ecosystems of the planet. In 2013 McKibben was awarded the Steward of Creation Award from the National Religious Coalition on Creation Care (NRCCC) for his more than thirty years of persevering and faithful advocacy.

Standing Up to the Herods in Ourselves and Society

Pope John Paul II named Bishop Jorge Mario Bergoglio the cardinal of Argentina in 2001. On the Feast of St. Joseph in 2013, the day when this cardinal was bestowed with the Fisherman's Ring in an outdoor Mass, Pope Francis followed John Paul II's model, calling on all human beings, especially Christians, to protect God's creation as St. Joseph protected Mary and Jesus. This was not a feel-good papal nosegay—it was an urgent call to see evil at work in the world and stand against it. "Tragically," Pope Francis said, "in every period of history there are 'Herods' who plot death, wreak havoc, and mar the countenance of men and women." It is our duty to be protectors of God's gifts, he said, and "whenever human beings fail to live up to this responsibility, whenever we fail to care for creation and for our brothers and sisters, the way is opened to destruction and hearts are hardened."

To prevent this end, Pope Francis urged everyone to wake up and do what they can to stop the destruction:

> Please, I would like to ask all those who have positions of responsibility in economic, political and social life, and all men and women of

goodwill: let us be "protectors" of creation, protectors of God's plan inscribed in nature, protectors of one another and of the environment. Let us not allow omens of destruction and death to accompany the advance of this world!

INAUGURATION MASS HOMILY, VATICAN CITY, MARCH 19, 2013

In this, Pope Francis imitated Christ, who cared for the poor, talked to strangers, drove the money changers out of the temple, raged at hypocrites, and favored praying in God's own places—the wilderness, deserts, gardens, on lakes and hills.

Pope Francis's Teachers in Caring for Creation

Pope Francis's direct predecessor, Pope Benedict XVI, had been dubbed "the green pope" because he installed donated solar panels onto the roof of the Vatican, retrofitted some of the pope vehicles for alternative fuels, and promoted the "Ten Commandments of the Environment" from the Pontifical Council for Justice and Peace. However, Benedict saw himself not as a pathfinder but one chosen to carry on the work of his saintly friend and predecessor. In all of his speeches about creation, Benedict directly quoted and echoed John Paul II, who had laid out the comprehensive framework of the theology of the earth's body, just as he had for that of the human body.

Today's Herods and Cold Evil

Like Pope Francis, John Paul II too had preached against the "Herods"— the corporate representatives, media personalities, reporters, government officials, and industries that were discrediting the warnings of the scientific community and promoting (or accepting) ecological destruction for profit. His 1997 alarm to the UN rings even more true today after decades of climate change denial:

The environment has often fallen prey to the interests of a few strong industrial groups, to the detriment of humanity as a whole, with the ensuing damage to the balance of the ecosystem, the health of the inhabitants and of future generations to come. . . .

ADDRESS TO UN CONFERENCE ON HEALTH AND THE ENVIRONMENT, MARCH 24, 1997

The fossil fuel industries and their executives, with their stranglehold on energy and their campaigns of misinformation and climate change denial, fit the pope's industry description—the Herods of today. This reprimand was only a fragment of John Paul II's assessment. In his global travels, he had witnessed that as personal and business morals dissolve, cold evil seeps in, resulting in slums, pollution, war, famine, poverty, public health disasters, sexual trafficking, depression, suicide, and domestic abuse: a culture of waste.

"Cold evil" is the term Jeremy Rifkin coined in his book *The European Dream* to describe decisions that prioritize profits, consumption, and control over the needs of people and nature. It is cold evil because the decision makers and consumers are distanced by time and space from those who are harmed. Therefore, individuals can choose actions that bring evil outcomes with cool, calm heads, justifying them as profitable, practical, efficient, or prudent. For example, the decision of a mining company to blast off the top of a mountain to extract the coal underneath ends up not only destroying the beauty and vitality of the ecosystem and human homes but polluting the water sources, spreading human birth defects, miscarriages, cancer deaths, allergies, and many other health problems among the local people because of the toxins released into their air and rivers. The public health statistics collected by Christians for the Mountains and other organizations in West Virginia show the enormous increase in health issues related to this type of mining. University researchers presented data to Congress that showed there was an excess of 4,332 deaths in the mountain removal areas in that state. This does not even take into account the negative effects on quality of life. This is cold evil. Not a great trade-off for limited-time jobs.

The banality of this evil ripples out in so many directions. No one considers that when they are turning on a light switch somewhere else in West Virginia or on the East Coast that they are supporting these destructive mountain blasts. Or adding to the mountains of coal burned, polluting the skies, adding to acid rain and mercury poisoning, and contributing to global warming. And consequently to loss of species of river fish and ocean coral turning white and dying as sea water gets warmer and more acidic... the cold evil spreads.

In various speeches and encyclicals, Pope John Paul II labeled these actions of cold evil variously as social evils, selfishness, sin, social sin, or structures of sin. He warned:

> In the *secularized modern age,* we are seeing the emergence of a twofold temptation: knowledge no longer understood as wisdom and contemplation but as power *over* nature...
>
> The other temptation is the unbridled exploitation of resources under the urge of unlimited profit-seeking, according to the capitalistic mentality typical of modern societies.
>
> ADDRESS TO UN CONFERENCE ON ENVIRONMENT AND HEALTH, MARCH 24, 1997

On a personal level, cold evil oozes out as denial or dismissal of the scientific or social realities. Consider stocks: many people view investing in stocks as an everyday decision simply involving the tabulations of profit trends, fees, and risk. Yet Christ told us that where our heart is, our treasure also shall be. Do our investments match our values? What of the companies that contribute to pollution, global warming, and huge executive remunerations for those laying off workers? John Paul II forced us all to see the connections.

> Destruction of the environment highlights consequences of decisions made by private interests that do not weigh the real conditions of human dignity. One finds prevalent an unbridled desire to accumulate personal wealth that prevents people from hearing the alarming cry of poverty of entire peoples.... the selfish quest for their own good fortune induces people to disregard the legitimate expectations of present and future generations.
>
> MESSAGE FOR WORLD TOURISM DAY, "SAVAGE TOURISM," JUNE 24, 2002

Accountability, responsibility, and love were the measuring sticks John Paul II demanded that we use to view all our actions, saying, essentially, "the buck stops here." Nature does its part, but we do more than our share. John Paul told us to step up:

> ...it is possible to say that *human behavior is sometimes the cause of serious ecological imbalance,* with particularly harmful and disastrous consequences in different countries and throughout the world. It suffices to mention armed conflict, the unbridled race for economic growth, inordinate use of resources, pollution of the atmosphere and water.
>
> ADDRESS TO PONTIFICAL ACADEMY OF SCIENCES, MARCH 12, 1999

When we invest, then, we can either invest to promote and heal the Kingdom of God or we can invest in that which undermines it at its very roots. John Paul counseled:

> It is therefore necessary to create lifestyles in which the quest for truth, beauty, goodness, and communion with others for the sake of common growth are the factors that determine consumer choices, savings, and investments....
>
> ENCYCLICAL, *CENTESIMUS ANNUS,* 1991

As we have all participated in getting the earth to this state and have lived off its systems, so Pope John Paul II said we *all must change*—from individuals to faith communities to local municipalities, from businesses and corporations to the media and artists, to states, nations, and international institutions and alliances.

> As one called to till and look after the garden of the world (cf. Gen. 2:15), man has a specific responsibility towards the environment in which he lives, towards the creation which God has put at the service of his personal dignity, of his life, not only for the present but also for future generations....
>
> The task of accepting and serving life involves everyone; and this task must be fulfilled above all towards life when it is at its weakest.
>
> ENCYCLICAL, *EVANGELIUM VITAE,* 1995

Answering the Call to Walk in St. Francis's Footsteps

John Paul II urged each of us to find in our own work and daily lives ways to live out our vocations of service to God, each other, and the whole of creation. He asked us to find ways to make courage and hope contagious, to transform our cultural norms, taboos, and collective consciences, not just because of the benefits that shall come back to us, but because God the Creator has asked us to do so and continues to ask us to be part of the co-creation of the world each day. St. John Paul II sends us his blessing across the years on this sacred and rewarding work:

> In your endeavor to preserve the healthiness of the environment, may the Lord enlighten and assist you.
>
> I commend your efforts to his bounty as our Father, rich in love for each one of his creatures, and I bless you all in his name.
>
> ADDRESS TO UN CONFERENCE ON ENVIRONMENT AND HEALTH, MARCH 24, 1997

PART II

THE ECOLOGICAL EMERGENCY AND MORAL CRISIS:
The Scriptural Call for Urgent Action

*How can we deny that today humanity is
experiencing an ecological emergency?*

WORLD DAY OF TOURISM, 2002 (SEPTEMBER), ISSUED JUNE 24, 2002

*It is the ecological question
ranging from the preservation of the natural habitats
of the different species of animals and of other forms of life,
to "human ecology" properly speaking
that finds in the Bible clear and strong ethical direction
leading to a solution that respects the great good of life,
of every life.*

ENCYCLICAL, *EVANGELIUM VITAE*, SECTION 42, MARCH 25, 1995

The troubled situation of this world of ours
at the dawn of the Third Millennium
has one advantage, if I may say so:
it makes us squarely face our responsibilities.

SPEECH TO DIPLOMATIC CORPS, JANUARY 10, 2002

The Ecological Emergency
Is a Moral and Faith Emergency

Each generation, in its time and place, is faced with some overarching moral challenges to which it must respond. Such is the case with us. St. John Paul II and scientists throughout the world warned for years that our planet was heading toward an ecological tipping point because of its overstressed systems.

By 2002, John Paul II was asking the world in semi disbelief: "How can we deny that today humanity is experiencing an ecological emergency?"

Now the pace of ecological degradation is accelerating. The ice caps are melting decades earlier than any of the computer projections predicted, unleashing seeping caches of buried methane—an even more concentrated greenhouse gas than carbon. This creates a feedback loop that multiplies the climate chaos. Widespread, unexpected weather events and natural disasters are becoming the norm in the news: hurricanes and wildfires devastating entire regions, tragic tornado sprees, droughts and draining aquifers, widespread floods, spreading deserts.

It's difficult to say something is not amiss. We've reached 400 parts per million of carbon in the atmosphere. This number may not mean much to nonscientists, but this is the highest percentage since the Pliocene Epoch, long before humans walked the earth. For life as we know it to be sustained, the highest carbon level our planetary systems can tolerate over the long term is only 350 parts per million, a mark we passed in the 1990s.

Every record-breaking heat wave, every freak snowstorm, every species on the brink of extinction reminds us that if we don't enact sweeping changes in how we interact with the earth soon, its ecological systems, pushed to their limits, will begin shutting down like organs in an ailing body.

In the summer of 2012, Bill McKibben tallied some of the newest numbers:

> June broke or tied 3,215 high temperature records across the United States.... That followed the warmest May on record for the Northern Hemisphere—the 327th consecutive month in which the temperature of the entire globe exceeded the 20th-century average... Saudi authorities reported that it had rained in Mecca despite a temperature of 109 degrees, the hottest downpour in the planet's history.

His frightening math goes on, but that's plenty. (You can find the recent figures on the U.S. National Oceanic and Atmospheric Administration's website, www.climate.gov, the NASA website, or that of *National Geographic*, among many other reputable sites on the Internet.) The burners are up so high on this planet that the oceans are not only warming but getting more and more acidic, since carbon mixed with water equals carbonic acid. This means we are slowly cooking the world's sea life in an acidic marinade. This is the same food chain that coastal people (and others), wildlife, and sea birds depend on. Add to this the fact that pollinators in the grain belts of the world can't function in super high heats, threatening the earth's breadbaskets. Expanding the problems, the practices of clear-cutting and slash-and-burn logging, plus the ensuing heat waves and drought, have set the stage for the wildfires scouring our planet of the forests that keep in the moisture, give us oxygen, and provide water to our underground aquifers and river springs. Fire is natural and helps forests stay healthy, but too much at the same time robs the ground of its shady cover and wildlife of its habitat. Deserts are taking the place of forests. Fresh, clean water is growing scarce. *Emergency* is not a strong enough word.

The trouble is, many of us either are exhausted by all this scientific news or consider such concerns controversial, based on whether we "believe" in climate change or not. We sometimes forget that the earth goes its own way and doesn't care if we believe in its processes or not. Yet even if you remember this, it's hard to have the continued emotional energy to keep paying attention—especially when your friends don't appear to be doing so. Environmental issues seem linked to the contentious world of politics, economics, and science—not to religion and morality. And it turns out, that's one of the biggest problems. And one of the biggest fallacies.

The Universal Moral Arguments

Even before John Paul II, writers, philosophers, religious leaders, and activists from diverse backgrounds had been laying out the compelling moral arguments for caring for the environment. In America, individuals like Emerson and Thoreau through to John Muir, Aldo Leopold, and Rachel Carson, to Carl Sagan, McKibben, and so many others had made the case for taking better care of the planet. When John Paul II arrived on the scene in 1978, though, he expanded the arguments beyond the issues of nature to include the human effects as part of nature: How could environmental damage *not* be a moral issue when it so increases suffering in the world, especially for the poor? How can it be moral or healthy to be extracting massive amounts of natural resources for things that are disposable, transforming beautiful and healthy habitats into wastelands and piles of garbage, while we suffer soul sickness from materialism? And how can it be acceptable to leave the world in far less habitable condition for our own children and future generations?

> Humanity has in its possession a gift that must be passed on to future generations, and, if possible, passed on in better condition.
>
> APOSTOLIC EXHORTATION ON THE VOCATION AND MISSION OF THE LAY FAITHFUL IN THE CHURCH AND THE WORLD, *CHRISTIFIDELES LAICI*, DECEMBER 30, 1988

Our only hope would be a radical reorientation of who we are, and who we are in this planet—an "ecological conversion" as he called it, and he described why and how.

Unfortunately religion and morality have bad reputations in the present national and international cultures, associated as they are with judgmental fundamentalist finger pointing, politics, and terrorism. Religious rhetoric and actions in the public forum have often been angry, mean-spirited, slanderous, and toxic. All of this undermines any validity to claims made in the name of religion or morality.

That's not what John Paul II's moral vision was about—not division, but dialogue, respect, and solidarity. He explained that ethics and religion are how we explore and understand the truths about the very origins and meanings of our existence and struggle to find the most healthy, harmonious, and freeing ways to live with one another on this earth—freedom *with* and *through* limits and God-given responsibilities. Pope John Paul II expressed it this way to the UN:

> For different cultures are but different ways of facing the question of the meaning of personal existence. And it is precisely here we find one source of the respect due to every culture and every nation: every culture is an effort to ponder the mystery of the world and in particular, of the human person: it is a way of giving expression to the transcendent dimension of human life. *The heart of every culture is its approach to the greatest of all mysteries: the mystery of God....*
>
> ADDRESS TO GENERAL ASSEMBLY OF THE UN IN HONOR OF
> ITS 50TH ANNIVERSARY, NEW YORK, OCTOBER 5, 1995

In addition to the greatest of all mysteries, what all cultures, nations, and religions share is the earth itself, upon which we all depend, and concern for the future of our offspring. These are the structural core of a global moral solidarity for ecological conversion and solidarity of action.

That is why Pope John Paul II was so insistent that we must stop the destructive rhetoric and get beyond our own biases and barriers to respect one another and be civil. Everyone knows that it is far easier to make progress on change when the people, products, and culture around you support it. If it were the norm to have white roofs in hot places and black ones in cool places, and solar panels on them all, we'd do it without thinking. If it were fashionable to bicycle to work and if cities had bike trails and lanes as a matter of course, more people would do it. If it were shameful and more expensive to buy products with excessive packaging or that were not built to last, fewer people would buy them. In general, if we lived in a society where caring was seen as a sign of strength instead of weakness, just imagine what change could be accomplished...and quickly.

Not a Political Issue, But One of Faith and Ethics

However, to see things in their proper perspective, Pope John Paul II felt we must acknowledge how faith in a Creator fits into the picture.

> The truth is that when people cut themselves off from God's plan for creation, they block out concern for their brothers and sisters, and respect for nature.
>
> However, there are reasons for hope. Many persons, aware of this problem, for some time have been studying ways to find a remedy. They are first of all concerned to recover the spiritual dimension of the relationship with creation, by rediscovering the mandate God originally entrusted to humanity (cf. Gen. 2:15).
>
> MESSAGE FOR THE 23RD WORLD DAY OF TOURISM, JUNE 24, 2002

In disentangling all the related threads for us, Pope John Paul II began where the earth began in Scripture. The Bible's first creation story made clear that care of this planetary garden was not a matter of political or personal choice but a duty at the very core of faith and existence—the first responsibility given to humans by the Creator. The pope explained:

> In the book of Genesis, where we find God's first self-revelation to humanity, there is a recurring refrain: *"And God saw that it was Good."*
>
> After creating the heavens, the sea, the earth and all it contains, God created man and woman. At this point the refrain changes markedly: And God saw everything He had made, and behold, *It was Very Good....*
>
> God entrusted the whole of creation to the man and woman, and only then—as we read—could He rest "from all his work."
>
> "PEACE WITH GOD THE CREATOR, PEACE WITH ALL OF CREATION," JANUARY 1, 1990

This is a mind-blowing concept—that a Creator mysterious and powerful enough to form a universe would put the future of life on earth in the hands of one creature species. This puts people into thoughtful partnership with God on a practical, day-to-day level.

> Created in the image of God, humanity has the right to make use of
> other created realities *but not to lord over nature, still less to ruin it. People*
> *are called to become God's coworkers in caring for creation.*
>
> ANGELUS, MARCH 24, 1996

Naturally, success in any partnership with a mysterious Creator demands
humility on the part of the creatures, to remain in awe of the Creator's
greater and more mysterious intents and the universe's infinitely complex
systems. The ecological ethicist Aldo Leopold, who died in 1948, had earlier
pressed the question of human humility:

> If there be, indeed, a special nobility inherent in the human race—a
> special cosmic value, distinctive from and superior to all other life—by
> what shall it be manifest?
>
> By a society decently respectful of its own and all other life,
> capable of inhabiting the earth without defiling it? Or by a society, like
> that of John Burroughs' potato bug, which exterminated the potato,
> and thereby exterminated itself? As one or the other shall we be
> judged in the "derisive silence of eternity."*
>
> *THE QUOTATION ON ETERNITY IS FROM "WALKING TOURS," BY ROBERT LOUIS STEVENSON;
> ALDO LEOPOLD, "SOME FUNDAMENTALS OF CONSERVATION IN THE SOUTHWEST,"
> *THE RIVER OF THE MOTHER OF GOD AND OTHER ESSAYS* (1991)

This was one of the conundrums for Leopold, who was raised a Lutheran
and married a Catholic. How could religious people who pledge love and
obedience to a Creator not sense an obligation to care for the works of that
Creator? How could humans of any religion believe that they know God's
purposes in the cosmos and that these purposes pertain only to this particular human species on this particular planet in this portion of universal time?

Religious hubris seems particularly incongruous in this century when
one views the awe-inspiring photographs of the vastness of the universe
from the Hubble Space Telescope. The "Sombrero Galaxy" (officially known
as M104) is 28 million light-years away from us and is estimated to be 50,000
light-years across and to have *800 billion suns.* How many planets do these
suns hold in orbit around them? This galaxy, smaller than our own Milky
Way, is only one of the 3,000 galaxies photographed in the range of the

Hubble telescope. The numbers are so staggering we can't even begin to fathom what they mean. How could we ever know how many billions of other planets may hold living ecosystems and intelligent life beloved by God? Or what purposes God has for this planet?

Clearly a deep humility and acknowledgment of ourselves as creatures with very limited vision is needed. John Paul II observed: "Only this humility before the grandeur and mystery of creation can save man from the ill-fated consequences of his own arrogance" (MESSAGE TO A MEETING FOR FRIENDSHIP AMONG PEOPLES AT RIMINI, ITALY, AUGUST 23, 2004).

Pope Francis has reiterated John Paul II's call for humility. He told a Jesuit interviewer, "We must be humble...God is always a surprise, so you never know where and how you will find him." He added: "Human self-understanding changes with time, and so also human consciousness deepens. Let us think of when slavery was accepted or the death penalty was allowed without any problem. So we grow in understanding of the truth."

And who was more humble than the saint who followed Christ so passionately that he tossed off all his clothes and riches in the plaza square to take on simple homespun and live with the poor? St. Francis gave us that example of how to humbly rejoice in all of God's creatures and in the messiness of life, where God is always waiting to surprise us.

Unfortunately, centuries of unchecked arrogance have led to many ecological missteps: dust bowls, extinct species, water and air pollution, famine. In 1967, medieval scholar Lynn Townsend White Jr. published an article in the journal *Science* titled "The Historical Roots of Our Ecological Crisis." He posited that present environmental woes result partially from Christianity's interpretation of the passage in Genesis when God gives humans "dominion" over nature (according to English translations). Especially during the Industrial Revolution, the word was often taken as permission, even a mandate, for many Christians to do anything they wanted with nature. This was intensified by the assumption that humans were made separate from the rest of nature. White, however, admitted that St. Francis was a notable exception.

Dominus, Not Domination

John Paul II corrected these scriptural misinterpretations. Ironically, that sometimes troublesome word "dominion" means precisely *the opposite* of the misinterpretation. The word "dominion" comes from the Latin *dominus,* to be of the Lord. Thus Genesis says we humans need to be working in the Garden of God, in the image and values of God, who said that all creation was good. So God entrusted the valued earth to humans' *care,* not domination or abuse. And the pope reminded us: "This is a task that humanity must carry out in respect for the divine image received, and, therefore, with intelligence and with love, assuming responsibility for the gifts that God has bestowed and continues to bestow" *(CHRISTIFIDELES LAICI,* DECEMBER 30, 1988).

That means in public policy, business, development, and our own personal lives, we must remember to view dominion through a nonselfish lens, as the pope makes clear in his encyclical on contemporary societal concerns:

> On the basis of this teaching, development *cannot* consist only in the use, dominion over, and indiscriminate possession of created things and the product of human industry, but rather in subordinating the possession, dominion, and use to man's divine likeness and to his vocation to immortality.
>
> ENCYCLICAL, *SOLLICITUDO REI SOCIALIS,* DECEMBER 30, 1987

That is why humility is so necessary to overcoming our selfish and limited understandings of nature and our role in it.

By honoring and blessing each piece of creation as it was being made *as good,* and the whole of it at the end as *very good,* God made priorities clear. Care for it all, each and every piece, and the whole as well. This is why science is a crucial part of this mandate. Without it, we don't really know *how* to care for the whole biota, since biological relationships are so intricate and complex. Scientists are dedicated to the process of learning and trying to accumulate understanding that can help us to be more ethical in our interactions, or dominion. Aldo Leopold offered this ecological wisdom in down-to-earth terms:

The last word in ignorance is the man who says of an animal or plant:
"What good is it?" If the land mechanism as a whole is good, then
every part is good, whether we understand it or not. If the biota, in
the course of aeons has built something we like but do not understand,
then who but a fool would discard seemingly useless parts? To keep
every cog and wheel is the first precaution of intelligent tinkering.

"THE ROUND RIVER," *A SAND COUNTY ALMANAC:
WITH OTHER ESSAYS ON CONSERVATION FROM ROUND RIVER*

Not surprisingly, Pope John Paul II told us to look to St. Francis as
a model of how to do this, how to honor every cog and wheel, for he
lived out "an example of genuine and deep respect for the integrity of
creation":

As a friend of the poor who was loved by God's creatures, St. Francis
invited all of creation—animals, plants, natural forces, even Brother
Sun and Sister Moon—to give honor and praise to the Lord. The poor
man of Assisi gives us striking witness that when we are at peace with
God we are better able to devote ourselves to building up that peace
with all creation, which is inseparable from peace among all peoples.

It is my hope that the inspiration of St. Francis will help us to
keep ever alive a sense of "fraternity" with all those good and beautiful
things which Almighty God has created. And may he remind us of our
serious obligation to respect and watch over them with care, in light of
that greater and higher fraternity that exists within the human family.

WORLD DAY OF PEACE MESSAGE, "PEACE WITH GOD THE CREATOR,
PEACE WITH ALL OF CREATION," JANUARY 1, 1990

The duty of kindness toward animals and all of creation was laid out in
the earth's structure from the very beginnings in Judeo-Christian Scripture,
and it is ever renewed with the dawn of each day.

A Servant Leader to All Species
in the Community of Creation

So how do we follow St. Francis in this? By following, as he did, Jesus's example of love and service. Jesus told his disciples, "If I, therefore, the master and teacher, have washed your feet, you must go and wash others' feet. I have given you a model to follow, so that as I have done for you, so you must do for others." That means the stranger as well as the friend, as Christ made clear in his story of the Good Samaritan. For St. Francis, strangers and "others" meant animals and plants as well as people. It is said he even befriended a threatening wolf in Gubbio, establishing a covenant of caring between the frightened villagers and the wild predator.

Christ, no doubt, also treated all parts of creation with care and respect, as Pope John Paul II told young people in Denver in 1993: "Jesus teaches us to see the Father's hand in the beauty of the lilies of the field, the birds of the air, the starry night, fields ripe for the harvest. "St. Francis, devoted to Jesus, followed his model, and preached:

> Not to hurt our humble brethren (animals) is our first duty to them,
> but to stop there is not enough. We have a higher mission—to be of
> service to them whenever they require it. If you have men who will
> exclude any of God's creatures from the shelter of compassion and pity,
> you will have men who will deal likewise with their fellow men.
>
> FOUND IN VARIOUS SOURCES ON THE SAINT, INCLUDING ST. FRANCIS
> FOUNDATION FOR ANIMALS: WWW.SAINTFRANCISFOUNDATION.COM

Echoes of St. Francis's words resound centuries later in Pope John Paul II's own counsel about animals:

> When man habitually loves and respects lower creatures, he will also learn
> to be more human with his equals. I am therefore happy to encourage
> and bless those who work to assure that, in the Franciscan spirit, animals,
> plants, and minerals be considered and treated as "brothers and sisters."
>
> POST-ANGELUS: GREETING TO PARTICIPANTS IN
> TERRA MATER [MOTHER EARTH] SEMINAR, OCTOBER 3, 1982

Our own devotion to our pets, livestock, and favorite wild birds and animals shows us that we have a natural understanding of this. Our houseplants, trees, gardens, and favorite places in nature hold special places in our hearts, demonstrating the deep emotional connection we have to other species. Scientist E. O. Wilson has a theory that this *biophilia*, or love of things alive and vital, is an instinct built into us. Though we can have repulsion or aversions to certain potentially dangerous forms of life, such as spiders and snakes (which can be poisonous), overall we retain a hardwired drive that makes us subconsciously seek out connections to the systems and species of the rest of life. Think about how many people choose to vacation near oceans, mountains, lakes, and rivers even when they say they don't like nature. Or those who become depressed when the sun doesn't come out for too many days in a row. You can see it so vividly when after days of cold and rain, people naturally burst out from indoors, seeking the sun.

There is some scientific and anecdotal evidence that biophilia is not just in humans toward other animals and nature, but also in other animals toward us. Pets' devotion to their human companions is an obvious example, but wild animals often show a curiosity and an affinity for watching humans as well, such as chickadees that follow hikers down a path, flitting and calling from tree to tree.

Scientists are often struck by incidents of what seems like biophilia from wild animals they study or work with. There are several incidents where individuals have found wild whales stranded in discarded fishing nets left floating in the water (part of the overwhelming problem of plastic wastes in the oceans). After these people have spent hours working to carefully free the whales, the animals have not simply swum away, glad to be free. In one case, the whale went to each of the divers separately to swim alongside them, allowing itself to be touched. In another, the whale swam alongside the boat for over a half hour, soaring up and plunging down in the water in what looked like a joyful dance of gratitude. Rescued loons do similar displays. When conservationist Lawrence Anthony, known as "the Elephant Whisperer," died, two herds of wild elephants inexplicably traveled days from different areas to gather at his home. They stayed a day or two, as if mourning one of their own, then returned to their home territories. These are the personal interpretations from the human witnesses, of course, but the animals' actions represent changes in behavior from their natural pattern of avoiding too much human contact out of fear.

Scriptural Biophilia

The Bible recognizes this natural affinity, as God built it into life. As God worked through the seven days of creation (later Biblical passages point out quite logically that days are not exactly the same measure of time for a Creator as for a creature: "with the Lord one day is like a thousand years and a thousand years like one day" (2 Peter 3:8)), God commissioned the earth to "bring forth" the animals and plants, which nicely fits with evolution. After the planetary stage was set, it was time for humans to appear on the scene. They were formed out of the earth, given the image and breath of God. Yet God didn't want this poor humanoid (not yet quite man or woman) to go sitting around all by its lonesome. So in the Bible's second story of creation, the Lord God presented humanity with all the animals *as friends*:

> The Lord God said: "It is not good for the *adam* to be alone. I will
> make a suitable partner for him." So the Lord God formed out of
> the ground various wild animals and various birds of the air, and he
> brought them to the man to see what he would call them...
>
> GENESIS 2:18–19

This naming of the animals was an act of great intimacy and connection. It is human nature that people tend to treat the things and people that we know the proper names of with greater respect and care, and those with which we are less familiar with more fear, apathy, or cruelty. This is why learning the names and features of wild animals and plants, and other peoples' customs and cultures, helps us fear them less and love them more. Obviously God recognized that the animals were a good start as company but not great for in-depth conversation or right for romantic procreation. This does not diminish, however, their original role as our companions, which God reminds us of when He establishes the co-covenant with all of life after the Flood:

"I am establishing my covenant with you and your descendants after you, and with every living creature with you, the birds, the domestic animals, and every animal of the earth…"

GENESIS 9:8-17

Humans need to be continually reminded of the inherent, God-created dignity and worth of animals because we often don't treat animals well. For instance, we have developed livestock industries that are dependent upon inhumanely caging massive numbers of animals in small areas in factory farms, then transporting and slaughtering them in great suffering. We conduct horrific experiments on animals in laboratories, some for cosmetic or theoretical purposes rather than medical necessity. We have led wild animals to extinction by hunting them or destroying their habitats with little consideration for their welfare.

John Paul II understood the medical necessity of working with animals but reminded the world that animals possess the divine spark of life, so human beings must love and feel solidarity with our smaller brethren. The Bible makes this clear by sometimes using the same Hebrew word for "soul" for both humans and animals and sometimes using a different one that suggests differences in nature and complexity:

> …we know that man was created "in the image and likeness of God" (1:26-27). Other texts, however, admit that the animals also have a vital breath or wind and that they received it from God. Under this aspect man, coming forth from the hands of God, appears in solidarity with all living beings. Thus Psalm 104 makes no distinction between men and animals when it says, addressing God the Creator: "These all look to you, to give them their food in due season. When you give to them, they gather it up" (vv. 27-28). The Psalmist then adds: "When you take away their breath, they die and return to their dust. When you send forth your Spirit, they are created; and you renew the face of the earth" (vv. 29-30).

GENERAL AUDIENCE, "THE CREATIVE ACTION OF THE DIVINE SPIRIT," JANUARY 7, 1990

So just as humans are made from the same physical matter as animals and plants—the "clay" of the earth (the same water, cellular structure, minerals, and much the same genetic material, as scientists note)—we also share some of the same soul matter, God's Spirit and breath of life.

We are therefore all part of an interconnected community of life, coming from the same spiritual and physical sources. Aldo Leopold postulated in his essay "The Land Ethic" that "ethics are possibly a kind of community-instinct-in-the-making." He explained that "the land ethic simply enlarges the boundaries of the community to include soils, waters, plants, and animals, or collectively: the land." Pope John Paul II's word for the land community was "creation." This makes even deeper sense when you consider the definition of ecology from Leopold:

> Ecology is the science of communities and the ecological conscience is therefore the ethics of community life.... An ecological conscience, then, is an affair of the mind as well as the heart.
>
> "THE ECOLOGICAL CONSCIENCE," *THE RIVER OF THE MOTHER OF GOD*

Considering this, it is not surprising that Pope John Paul II consistently favored the term "ecology" over the less inclusive "environment." He also highlighted that to value all life doesn't mean to erase any differences, preferences, or priorities. It doesn't mean that God is calling us to value the life of a sparrow the same as the life of a child. That's where "dominion" comes in. It means seeing the full picture of creation, looking ahead, caring for all, and avoiding the antagonistic positioning of one against the other in God's community.

Christ in All Parts of the Community of Creation

This necessity to avoid antagonisms and care for all the animals, plants, and minerals of God's cosmos is not just an Old Testament concept. For in the very beginning of the Gospel of John, the evangelist proclaims that *Christ is in all of Creation.* "In the beginning was the Word, and the Word was with God, and the Word was God.... All things came to be through Him, and without Him, nothing came to be." So Christians are called particularly to recognize the Divine energy of God flowing through each of the pieces of creation, even the nonliving ones, and the system as a whole. John Paul II notes:

There is a growing threat to the environment of humanity, to vegetation, animals, water, and air. Sacred Scripture hands on the image of Cain who rejected his responsibility: "Am I my brother's keeper?"

The Bible shows the human person *is* his brother's keeper and the guardian of creation that has been entrusted to him.

ADDRESS, REPRESENTATIVES OF SCIENCE, ART, AND JOURNALISM, SALZBURG, AUSTRIA, JUNE 26, 1988

Explaining this responsibility to the community of creation to American youth, John Paul used a Gospel metaphor Jesus chose for Himself:

Man is called to be the good shepherd of the environment.... This is the task he was given long ago and which the human family has assumed not without success through its history, until recently when man himself has become the destroyer of his natural environment.

WORLD DAY OF YOUTH MASS HOMILY, CHERRY CREEK CAMPGROUND, DENVER, AUGUST 14, 1993

Dominus as Our Daily Ecological Vocation

Obviously, from a faith perspective then, taking care of this earth's garden is not a bit of extra-credit homework or moonlighting to do on the side, separate from our main jobs in life, but rather the definition that sets the framework for all living. The pope explained:

Awareness that man's work is a participation in God's activity ought to permeate...even the most ordinary everyday activities.

For, while providing the substance of life for themselves and their families, men and women are performing their activities in a way which appropriately benefits society. They can justly consider that by their labor they are unfolding the Creator's work, consulting the advantages of their brothers and sisters, and contributing by their personal industry to the realization in history of the divine plan.

ENCYCLICAL, *LABORE EXERCEN* (ON HUMAN LABOR), SEPTEMBER 14, 1981

John Paul II referred to this everyday awareness as an "ecological vocation." "Nature" is not separate from us in context of creation; wild nature and human society are part of each other. All of our human industries of energy production, farming, mining, logging, etc., depend upon natural resources, as do our everyday activities of eating, drinking, getting dressed, washing, and so on. So all elements of human society are mere subsets of the whole of nature on this planet and within the cosmos. Our ecological vocation is equal opportunity. It is not based only on religious beliefs but on the reality of how the natural world and spiritual world are put together—how very interdependent we all are with one another, the earth, and the solar system. We are but a pinprick on the fabric of the earth, which is a mere swatch in the cosmic quilt. John Paul II mused:

> Before the wonder of divine providence manifested in creation and in history, the human creature feels very small. At the same time, the human creature recognizes that he is the one who receives the message of love inviting him to responsibility. Human beings are appointed by God as stewards of the earth to cultivate and protect it. From this fact there comes what we might call their "ecological vocation," which in our time has become more urgent than ever.
>
> ANGELUS, CASTEL GANDOLFO, AUGUST 25, 2002

Our ways of living and making a living, the pope pointed out, are how we act out our co-creatorship, our dominion, our ecological vocation, our work in the garden—they are all one and the same. Our lives and ways of making a living are acts of creation every day, because we are thinking and making choices that affect others and continually shift, change, or create larger outcomes. Humans are specially designed for this task of being a coworker with God, which the pope explains with a concept from his native Poland:

> The opening page of the Bible presents God as a kind of exemplar of everyone who produces a work: the human craftsman mirrors the image of God as Creator. This relationship is particularly clear in the Polish language because of the lexical link between the words *stwórca* (creator) and *twórca* (craftsman)....

God therefore called man into existence, committing to him the craftsman's task. Through his "artistic creativity," man appears more than ever "in the image of God," and he accomplishes this task above all in shaping the wondrous "material" of his own humanity and then exercising creative dominion over the universe which surrounds him. With loving regard, the divine Artist passes on to the human artist a spark of his own surpassing wisdom, calling him to share in his creative power.

LETTER TO ARTISTS, APRIL 4, 1999

As one can see, this ecological vocation is not meant to be drudgery (though of course sometimes it can feel like it), but a gift. It's a privileged and discerning position, a sacred calling filled with daily decisions, labor, satisfaction, inspiration, ingenuity, and power that we can use well or poorly. Just think of what it feels like to plant your garden or choose that stock or stand up for what you believe in the realm of public policy.

Aldo Leopold mused wryly about one of his interactions as a co-creator on the denuded land he purchased and lovingly worked back to health:

Acts of creation are ordinarily reserved for gods and poets, but humbler folk may circumvent this restriction if they know how. To plant a pine, for example, one need be neither god nor poet; one need only own a good shovel. By virtue of this curious loophole in the rules, any clodhopper may say: Let there be a tree—and there will be one.

If his back be strong and his shovel sharp, there may eventually be ten thousand. And in the seventh year he may lean upon his shovel, and look upon his trees, and find them good.

"PINES ABOVE THE SNOW," *A SAND COUNTY ALMANAC*, AND *SKETCHES HERE AND THERE*

Barriers to Good Co-Creatorship

To be a helpful co-creator, one must have a sense of the interconnections of the universe and the affects of one's actions. A poet, John Paul II acknowledged people's lack of awareness and blatant misuse:

> The profound sense that the earth is "suffering" is…shared by those who do not profess our faith in God. Indeed, the increasing devastation of the world of nature is apparent to all. It results from the behavior of people who show a callous disregard for the hidden yet perceivable requirements of the order and harmony that govern nature itself.
>
> WORLD MESSAGE OF PEACE, JANUARY 1, 1990

Part of the reason we don't perceive this order and harmony is we all have an inherently limited view as human beings, creatures rather than the ultimate Creator. We cannot see the past, future, or all the present interconnections and perspectives. We also have a built-in self orientation: the scriptural original sin, our natural tendency to proudly favor our perspectives and needs over those of God and other people. It is one of our key ecological and spiritual challenges: to inform ourselves beyond our own views and act beyond our own needs and wants. John Paul II told young people:

> Selfishness makes people deaf and dumb; love opens eyes and ears, enabling people to make that original and irreplaceable contribution which—together with the thousands of deeds of so many brothers and sisters, often distant and unknown—converges to form the mosaic of charity that can change the tide of history.
>
> 11TH WORLD DAY OF YOUTH, NOVEMBER 25, 1995

Property Rights Are Responsibility Rights

The other scriptural reorientation needed involves property. Part of the problem for the "developed" countries is a kind of built-in cultural selfishness. It is based on slicing the land into little bits and selling it off. Ecologically, that makes it hard to remember the needs of the land or the whole community. Owners often focus on their property rights, forgetting those of their neighbors, wild animals and plants, and the system as a whole. Things get even worse when natural resources—such as minerals, natural gas, oil, coal, and water—are used for profit as if humans had personally created this business "capital."

Again it all goes back to the Garden of Eden. In the Old Testament (or Torah for Jewish people), we are reminded that God never signed away property rights to humans but merely gave us use. Since we are prone to forgetting, God made it abundantly clear that we do not *own* the land we live on in Leviticus 25:23: "The land shall not be sold in perpetuity; for the land is mine, and you are but aliens who have become my tenants." Our property rights are, in fact, really responsibility rights as God's stewards.

Frankly, this is not an easy concept. To get into moral alignment with God and with nature, John Paul II said that we'll have to think more like servants in regard to property:

> To be humble regarding our idea of ownership and to be open to the demands of solidarity... We have not been entrusted with unlimited power over creation, we are only stewards of the common heritage.
>
> COMMON DECLARATION ON THE ENVIRONMENT OF POPE JOHN PAUL II AND THE ECUMENICAL PATRIARCH BARTHOLOMEW I, ROME-VENICE, JUNE 10, 2002

That makes one pause and think: What are the faith limits of my property rights? Am I looking at my property as a servant entrusted with its care? In contemporary society, few people or businesses (and certainly few legal or political systems) take this approach. John Paul II had no time for this. "*The earth belongs to God!* It must therefore be treated according to his law" (JUBILEE OF THE AGRICULTURAL WORLD ADDRESS, VATICAN, NOVEMBER 11, 2000).

He said further:

> The land is a gift of God for the benefit of all.... It is not admissible to use this gift in such a manner that the benefits it produces serve only a limited number of people, while the others—the vast majority—are excluded from the benefits which the land yields. A... challenge is, therefore, presented to those that own or control the land.
>
> ADDRESS, PROPRIETORS AND WORKERS, BACOLOD CITY, PHILIPPINES; FEBRUARY 20, 1981

In the Old Testament, God so wanted to emphasize this Creator-tenant relationship that he designated every seventh year as a sabbatical year for the land, animals, and people for rest and restoration, and a chance to rely solely on God's abundance, and faith, for a living. Every fiftieth year was the Jubilee Year, dedicated to spiritually remembering who we are and where we have come from, celebrating the generosity of God. During this year, all were to pause in rest to reorient and re-create their spirits in God, returning all lands for a year to the original families given the land after the 40 years in the desert. It was a year dedicated to returning to one's roots in God, weeding out injustice and serving the needy, realigning all social structures and systems of abuse that keep people, animals, and the land itself from health and fertility. This required both repentance for wrongs and reconciliation between warring parties. Catholics honor these Jubilee Years of equality, freedom, reorientation, and renewal.

During the last Jubilee Year, at the turn of the new millennium, Pope John Paul II emphasized the meaning of these restorative actions and the necessity of sharing and protecting the common good to balance individual ownership:

> It was a common conviction, in fact, that *to God alone, as Creator, belonged the...lordship over all Creation and over the earth in particular* (cf. Lev. 25:23). If, in his Providence, God had given the earth to humanity that meant that he had given it to *everyone*.
>
> Therefore, the riches of Creation were to be considered as a common good of the whole of humanity. Those who possessed these goods as personal property were really only stewards, ministers charged with working in the name of God, who remains the sole owner in the full sense....
>
> APOSTOLIC LETTER, *TERTIO MILLENNIO ADVENIENTE*, NOVEMBER 10, 1994

Resource Reform Needed for More God-Oriented Ownership

In 1997, in preparation for the Jubilee, John Paul II spoke to the UN Conference on Health and the Environment about two particular temptations to abuse our dominion and ownership that need to be overcome:

> In the *secularized modern age* we are seeing the emergence of a twofold temptation: a concept of knowledge no longer understood as wisdom

and contemplation, but as power over nature, which is consequently regarded as an object to be conquered.

The other temptation is the unbridled exploitation of resources under the urge of unlimited profit-seeking, according to the capitalistic mentality typical of modern societies.

MARCH 24, 1997

The pope explained later how these concepts of the land and Jubilee apply to contemporary use of property and natural resources:

This principle does not delegitimize private property; instead it broadens the understanding and management of private property to embrace its indispensable social function, to the advantage of the common good and in particular the good of society's weakest members....

VATICAN, DECEMBER 8, 1999

So what would that mean? Society's weakest members are its lowest-paid workers and the poor in the community (as well as other species). According to John Paul II's encyclical on work and proper use of ownership, *Laborem Exercens*, the workers should not be treated vastly differently from those who own the business or run it on the board of directors. Leaders of the Occupy Wall Street movement probably took moral tips from this document and also from his encyclical "On Social Concerns," *Sollicitudo Rei Socialis:*

One of the greatest injustices in the contemporary world consists precisely in this: that the ones who possess much are relatively few and those who possess almost nothing are many. It is the injustice of the poor distribution of the goods and services originally intended for all.

This then is the picture: there are some people—the few who possess much—who do not really succeed in "being" because, through a reversal of the hierarchy of values, they are hindered by the cult of "having"; and there are others—the many who have little or nothing—*who do not succeed in realizing their basic human vocation because they are deprived of essential goods.*

ENCYCLICAL, *SOLLICITUDO REI SOCIALIS*, DECEMBER 30, 1987

Supporting this, Pope John Paul II stated in *Laborem Exercens* that any use of any natural resource, land, or property (capital) "whether in the form of private ownership or in the form of public or collective ownership" should *serve* workers and the ability for dignified jobs to continue so that the goods or profits get universally distributed throughout the community. He proposed that proper business structures to accomplish this justly would be "joint ownership of the means of work, sharing by the workers in the management and/or profits of businesses, so-called shareholding by labor, etc." These business structures are not just for companies in developed countries but even more for multinationals and businesses in areas that do not have strong workers' rights and environmental laws in place, such as "the so-called Third World...in places of the colonial territories of the past" (ENCYCLICAL, *LABOREM EXERCENS*, SEPTEMBER 14, 1981).

Pope John Paul II did not favor socialism over capitalism, but rather responsible economic systems, such as limited capitalism, that encourage individual freedom, enterprise, and creativity, alongside moral responsibility, service, sharing, caring for the common good, and leaving no one unseated at the table of creation or merely eating its scraps. Catholic doctrine and all the modern popes have emphasized this same position. Pope Francis, however, has come out even more vocally in this fight, criticizing all who ignore the hungry and thirsty around the world.

God's business ethic of service does not just benefit the workers, but the poor who live near the places where natural resources are extracted. This ethic asks: do workers benefit from the business in terms of well-paying, safe, satisfying jobs, or are they paid poorly, work in dangerous conditions, and have their water and air polluted? Does the money stay in the community and help build up health, education, and the arts, improving the quality of life, or does it go to owners and stockholders in a different country? Does the business leave the community better or worse? John Paul II made it absolutely clear how we must think about these questions of proper use according to the values of God in an ecumenical address in Toronto. He said that united in the name of Christ, all Christians need to work within this religious and moral framework and interdenominational collaboration:

the needs of the poor must take priority over the desires of the rich; the rights of workers over the maximization of profits; the preservation of the environment over uncontrolled industrial expansion; production to meet social needs over production for military purposes.

APOSTOLIC VISIT TO CANADA, ANGLICAN CHURCH OF SAINT PAUL, SEPTEMBER 14, 1984

For the Jubilee Year, John Paul II proposed an immediate practical international reform to more carefully fit the role of people as tenants on God's lands to serve the whole community and common good. He advocated that the World Bank and international lenders forgive the crushing debts of developing countries. These in-debt nations have few ways to pay the money back without harmfully slashing their health and education systems and allowing multinational companies to exploit their natural resources, ruin their environments, and enslave their people in low wages and poor working conditions. These are all practices that developed countries would not allow in their backyards—a form of contemporary imperialism and racism, which John Paul stated was absolutely intolerable and unjust. When speaking with President George W. Bush, Pope John Paul II reiterated these same points:

> Respect for nature by everyone, a policy of openness to immigrants, the cancellation or significant reduction of the debt of poorer nations, the promotion of peace through dialogue and negotiation, the primacy of the rule of law: these are the priorities the leaders of the developed nations cannot disregard.

ADDRESS TO PRESIDENT GEORGE W. BUSH AT CASTEL GANDOLFO, JULY 23, 2001

This counsel of sharing the goods of creation is not new, nor is it going away. The early Church, with the apostles and new disciples, tried to imitate the designs of the Garden of Eden in its communal structure. And for centuries, Christians were forbidden to be bankers to help them avoid the temptation of forgetting to what service money must be used. Pope Francis told ambassadors that the world needs to reorient itself: "Money has to serve, not rule!" He observed of the world recession:

While the income of a minority is increasing exponentially, that of the majority is crumbling. This imbalance results from ideologies which uphold the absolute autonomy of markets and financial speculation, and thus deny the right of control to States, which are themselves charged with providing for the common good. A new, invisible and at times virtual, tyranny is established, one which unilaterally and irremediably imposes its own laws and rules.

Moreover, indebtedness and credit distance countries from their real economy and citizens from their real buying power. Added to this, as if it were needed, is widespread corruption and selfish fiscal evasion which have taken on worldwide dimensions. . . .

The will for power and possessions has become limitless. Concealed behind this attitude is a rejection of ethics, a rejection of God.

ADDRESS TO NEW PONTIFICAL AMBASSADORS, VATICAN, MAY 16, 2013

All of this is due to a skewed understanding of ownership, dominion, and the purpose of creation's resources. Individual stockholders also participate in this "cold evil" process. (It's analogous to a slum landlord situation, where the owners are absent and yet absorb the resources of the land for profit.) Popes Francis, Benedict XVI, and John Paul II spoke out against this and called for reform.

Pope Francis posed the challenge to bankers, corporations, politicians, and economists of realigning their priorities to be more in tune with the designs and harmony of creation:

And why should they not turn to God to draw inspiration from his designs? In this way, a new political and economic mindset would arise that would help to transform the absolute dichotomy between the economic and social spheres into a healthy symbiosis.

ADDRESS TO NEW AMBASSADORS, MAY 16, 2013

In the Jubilee, John Paul II noted that the realignment *must* be accomplished now, within the next fifty years in God's global garden because of the ecological crisis that "is making vast areas of our planet uninhabitable." He perceived it as God's invitation to us for a new, better life, one with God's aims for the common good coming first:

> The Jubilee is a further summons to conversion of heart through a change of life. It is a reminder to all that they should give absolute importance neither to the goods of the earth, since these are not God, nor to man's domination or claim to domination, since the earth belongs to God and to him alone: "the earth is mine and you are strangers and sojourners with me" (Lev. 25:23). May this year of grace touch the hearts of those who hold in their hands the fate of the world's peoples!
>
> BULL OF INDICTION OF THE GREAT JUBILEE OF THE YEAR 2000, *INCARNATIONIS MYSTERIUM*, SAINT PETER'S, ROME, NOVEMBER 29, 1998

Consequences of Irresponsible Dominion

What happens if we do not change our lives and ways of considering care of others, animals, property and money? No act of humanity or in nature is without effects or consequences. John Paul II reminded us that having the powers of God-given intelligence and ingenuity (and a shovel) "do not make man the absolute arbiter of the earth's governance, but the Creator's 'co-worker'—a stupendous mission—but one also marked by precise boundaries that can never be transgressed with impunity" (ADDRESS TO HONOR THE JUBILEE OF THE AGRICULTURAL WORLD, VATICAN, NOVEMBER 11, 2000).

Impunity is a scary concept. John Paul II explained that when we transgress against the natural order of the universe, breaking a system or removing key pieces, the consequences are spiritual, moral, *and* physical disturbances and dysfunction. He reminded us of what happened with the first man and woman and their use of the garden:

> Made in the image and likeness of God, Adam and Eve were to have exercised their dominion over the earth (Gen. 1:28) with wisdom and love. Instead, they destroyed the existing harmony *by deliberately going against the Creator's Plan,* that is, by choosing to sin. This resulted not only in man's alienation from himself, in death and fratricide, but also in the earth's "rebellion" against him....

> When man turns his back on the Creator's plan, he provokes a disorder which has inevitable repercussions on the rest of the created order. If man is not at peace with God, then earth itself cannot be at peace: "Therefore the land mourns and all who dwell in it languish, and also the beasts of the field and the birds of the air and even the fish of the sea are taken away" (Hos. 4:3).
>
> WORLD DAY OF PEACE MESSAGE, "PEACE WITH GOD THE CREATOR, PEACE WITH ALL OF CREATION," 1990

Pope John Paul II then cited many contemporary examples, including the greenhouse effect that "has now reached crisis proportions as a consequence of industrial growth, massive urban concentrations, and vastly increased energy needs"; depletion of the ozone layer; along with "industrial waste, the burning of fossil fuels, unrestricted deforestation, the use of certain types of herbicides, coolants, and propellants." Scientists measure the resulting natural impunity or disorder in terms of loss of key functions, species, fertility, flexibility, adaptability, and resilience of systems. The pope included the additional related consequences of environmental destruction and violence on people—poverty, refugees, disease, war—which then lead to more destruction and violence against the environment in a deadly feedback cycle.

John Paul II (and Christianity) taught that the consequences of sin are not only physical deaths but spiritual and moral deaths: death to harmony, death to our inner sense of peace and meaningfulness, death to good relationships with God and others, death to quality of life and community cohesiveness, death to species. These deaths, like the biological ones, are often gradual, little gradations of change that we don't notice within ourselves or our world, like the frog in a pot of water, slowly being heated until it dies. The poor frog doesn't jump out to save itself because the heat change is too gradual for it to notice—like global climate shifts, topsoil loss, and water and air pollution.

As for impunity after death? Each religion and each person conceives of concepts of the afterlife and punishment for earthly wrongs differently because it is all Mystery, veiled from view. But biblical Scripture and the Church teach that there is a hell of painful consequences for the sins of selfishness, destruction, and injustice that erode our souls and cause suffering to others, that alienate us from the free-flowing love of God. It's only natural. No act of love, or unkindness, is lost in the universe. Each has ripple effects and consequences. Why would it be different with our souls? Yet how do they work with God's own justice, mercy, second chances, and reparation?

This is where mercy and redemption come in. God's ever-flowing mercy and opportunities for renewal and reconstruction are offered to us in this world, and likely, the next. Only the Author of Life knows for sure the full consequences of our actions and how our lives shall offer judgment for and against us. Even Christ said that he didn't know, for that knowledge was left to his Father, the Creator (Mark 10:40). Which comes back to the concept of humility, and paying attention to God's mandates and designs up front, in this life.

To assist Jews and Christians in understanding the scriptural call to faith-filled care of creation, a committee of interfaith biblical and theological scholars worked together to identify in the Bible passages associated with caring for the earth. More than two thousand such passages are highlighted in green ink in a new Bible, *The Green Bible*, whose introduction begins with St. Francis's "Canticle of the Creatures." It also includes Pope John Paul II's 1990 World Day of Peace Message, "Peace with God the Creator, Peace with All of Creation," as well as contributions from significant others, such as Archbishop Desmond Tutu, J. Matthew Sleeth, M.D., of Blessed Earth, scholar Calvin B. DeWitt, and Rabbi Ellen Bernstein, who offers "Creation Theology: A Jewish Perspective." All the opening essays and the Bible itself have the same message: we are called to care for God's creation, and we must begin to act.

John Paul II told us:

> It is good that people remember that they find themselves in a "flow-erbed" of the immense universe, created for them by God. It is important for people to realize that neither they, nor the matters which they so frantically pursue, are "everything." Only God is "everything," and in the end, everyone will have to give an accounting of themselves to Him.

ADDRESS TO REPRESENTATIVES OF WORLD RELIGIONS, DAY OF
PRAYER OF PEACE IN THE WORLD, ASSISI, ITALY, JANUARY 24, 2002

*The seriousness of the ecological issue
lays bare the depth of humanity's moral crisis.
If an appreciation of the value of the human person
and of human life is lacking,
we will also lose interest in others and in the earth itself.*

WORLD DAY OF PEACE MESSAGE, 1990

The Gospel of Life
vs. the Culture of Waste and Death

In the 1980s and 1990s, the smiling Pope John Paul II epitomized the world's greatest pro-life advocate. Yet how many people considered him an outspoken champion of tropical rainforests, clean rivers, the humane treatment of animals, and the education of girls? The pope's pro-life vision broke out of any boundaries: "*life must be handled with care*, including animal life and all of animate and inanimate nature" (ADDRESS, REPRESENTATIVES OF SCIENCE AND ART, VIENNA, AUSTRIA, SEPTEMBER 13, 1983).

Violence, domination, disrespect, and despair were John Paul II's enemies. He fought them at every turn in his struggle to lift up all forms of life to care for, and honor, the dignity and beloved nature of each human. In his encyclical *Evangelium Vitae*, or "The Gospel of Life," he wrote:

> To choose life involves rejecting every form of violence: the violence of poverty and hunger, which oppresses so many human beings; the violence of armed conflict, which does not resolve, but only increases divisions and tensions; the violence of particularly abhorrent weapons, such as anti-personnel mines; the violence of drug trafficking; the violence of racism; and *the violence of mindless damage to the natural environment.*

The word "mindless" is key to the damage. In contemporary society, people do not think twice about making drastic changes to the natural world for things such as construction or mining or lumbering. Nor do they recognize that actions that slice, blow up, pollute, or spoil a habitat, air, water, or soil are a form of violence. Aldo Leopold explains why:

> Land, then, is not merely soil: it is a fountain of energy flowing through a circuit of soils, plants, and animals. . . . When a change occurs in one part of the circuit, many other parts must adjust themselves to it. . . .
>
> The combined evidence of history and ecology seems to support one general deduction: the less violent the man-made changes, the greater the probability of successful readjustment in the [energy] pyramid. Violence, in turn, would seem to vary with human population density; a dense population requires a more violent conversion of the land. . . .
>
> Can the violence be reduced? I think it can be, and that most of the present dissensions among conservationists may be regarded as the first gropings toward a nonviolent land use.
>
> "THE LAND ETHIC," *A SAND COUNTY ALMANAC*

Leopold believed that if cultures could come to value the whole biota (which includes humans) instead of just bits of it (such as only human endeavors, certain game species, or pieces of landscape), then progress could be made. He admitted in 1938: "Our tools are better than we are, and they grow faster than we do. They suffice to crack the atom, to command the tides. But they do not suffice for the oldest task in human history: to live on a piece of land without spoiling it."

The art of living as nonviolently as possible on the land is not one that most humans have yet learned as individuals or cultures, but John Paul II argued that this is *exactly what God has been calling us to do, from the very beginning*. When he saw some of the astounding photographs taken of earth from the moon and the satellites, John Paul II mused:

> For the first time, we see ourselves from outside—from the moon—and we realize that we must be more responsible for ourselves, our neighbors, our institutions and our planet, whatever may be our nation, religion or political stance.
>
> ADDRESS TO VATICAN OBSERVATORY CONFERENCE ON COSMOLOGY, JULY 6, 1985

Violence Against Nature Is Violence Against People

John Paul II observed that at the heart of the environmental violence wrought by humans lies a lack of respect for life, a reduction of living beings and natural systems to things to be manipulated for money:

> A lack of *respect for life* is evident in many patterns of environmental pollution. Often, the interests of production prevail over concern for the dignity of workers, while economic interests take priority over the good of individuals and even entire peoples. In these cases, pollution or environmental destruction is the result of an unnatural and reductionist vision which at times leads to a genuine contempt for man.
>
> On another level, delicate ecological balances are upset by the uncontrolled destruction of animal and plant life or by a reckless exploitation of natural resources.
>
> WORLD DAY MESSAGE OF PEACE, 1990

Satellite images over time show the truth of this, the encroaching desertification that is happening across the continents. Up close, talking to the victims, as the pope did in places like the Sahel belt in Africa, one can get a feel for the human suffering from global warming's intensification of droughts, heat waves, wildfires, floods, mudslides, hurricanes, tornadoes, blizzards, and famines. This is violence on a massive scale.

And violence begets violence. In a study published in August 2013 in the journal *Science,* an examination of 60 studies from 27 countries show an increase in violence and personal crime as the temperatures rise. Marshall Burke, the colead author of the study from University of California-Berkley, noted:

> We found that a one standard deviation shift towards hotter conditions causes the likelihood of personal violence to rise 4% and intergroup conflict to rise 14%.

The planet's poor (along with wild species) *always* bear the first and most devastating effects of ecological violence—the degradation and disasters—especially from climate chaos. They have the least protection from crime and the fewest resources to change with the times. They end up the refugees, the casualties of wars over resources, the long-term sufferers from chemical weapons and landmines, the inhabitants of toxic dumping areas, the drinkers of polluted water and breathers of dangerous air. They become the world's homeless, refugees always asking to be let into someone else's space, seeking clean water and food and shelter, fighting to keep their dignity and cultural identity. This is the connection and moral concept that the world at large was just not admitting, and still doesn't. Yet it was these, the least of our brothers and sisters, whom the pope visited, and in them, he saw the true meaning of climate change and ecological violence.

In 1999, as the pope prepared the globe for the new millennium, he appealed once more for a rejection of all forms of violence and then described how using a barometer of respect for life and an awareness of people as an essential part of creation could guide the world to greater care of all of nature:

> The world's present and future depend on the safeguarding of creation, because of the endless interdependence between human beings and their environment. Placing human well-being at the center of concern for the environment is actually the surest way of safeguarding creation; this in fact stimulates the responsibility of the individual with regard to natural resources and their judicious use.
>
> WORLD DAY OF PEACE MESSAGE, JANUARY 1, 1999

As the environmental movement has found, efforts to preserve environments and species that do *not* consider the needs of the local people generally fail because they are not sustainable. In the same way, however, efforts that prioritize human wants and greed over the health of the land, water, air, and other species are doomed to failure too, as humans are a part of nature and dependent on the sustenance of ecological systems. Because of the interdependence of humans as part of nature, there are ecological dimensions to war, education, social services and poverty, reproductive policies, farming, economics and trade, corporate laws, and so many other social issues.

Visiting his homeland in 1999, John Paul II returned to this theme: "It appears that what is most dangerous for creation and for man is lack of respect for the laws of nature and the disappearance of a sense of the value of life" (LITURGY OF THE WORD HOMILY IN ZAMOŚĆ, POLAND, JUNE 12, 1999).

Luther Standing Bear, a Lakota Sioux of the late nineteenth and early Twentieth Century, stitched up the premise simply and succinctly: "Man's heart away from nature becomes hard; lack of respect for growing, living things soon led to lack of respect for humans too."

And vice versa.

Aldo Leopold also linked respect for people and nature under the same umbrella of "life." Just after the end of World War I, he wrote:

> The truth is, that in spite of all religion and philosophy, mankind has never acquired any real respect for the one thing in the Universe that is worth most to Mankind—namely Life. He has not even respect for himself, as witness the thousand wars in which he has jovially slain the earth's best. Still less has he any respect for the other species of animals... The trouble is that man's intellect has developed much faster than his morals.

Realigning a Disordered Human Ecology

To address our pressing ecological crises, John Paul II explained how our human ecology (which includes beliefs, thinking, values, habits, choices, relationships, and possessions) needs to be reoriented toward God's view of life's goodness:

> At stake, then, is not only *a "physical" ecology*, which is concerned with safeguarding the habitat of the various living beings, but also *a "human" ecology*, which is concerned with the source of existence of all creatures and which by protecting the fundamental good of life in *all* its manifestations, prepares for future generations an environment more in conformity with the Creator's plan.
>
> GENERAL AUDIENCE, JANUARY 17, 2001

A disordered human ecology has repercussions at all levels of community. For instance, it is the same lack of respect for life that fosters environmental destruction that also allows an unborn baby to be seen as the property of the mother covered by the right of privacy; or women as the property of their husbands and fathers; or children as the property of their parents; or girls and boys as sexual slaves and toys; or laborers as slaves or near slaves; or animals as unfeeling property of their companions or livestock operations; or the those with less affluence and power as disposable people who must endure the onslaughts of environmental pollution while the affluent remain immune. It's the concept that people and other living beings can be treated as things, as possessions, as movable pegs—whereas money and property should be protected.

A complete realignment is required, according to John Paul II:

> Is it really possible to oppose the destruction of the environment while allowing, in the name of comfort and convenience, the slaughter of the unborn and the procured death of the elderly and the infirm, and the carrying out, in the name of progress, of unacceptable interventions and forms of experimentation at the very beginning of human life?
>
> When the good of science or economic interests prevail over the good of the person, and ultimately of whole societies, environmental destruction is a sign of a real contempt for people.
>
> LITURGY OF THE WORD HOMILY, ZAMOŚĆ, POLAND, JUNE 12, 1999

Consumerism Creates a Lack of Respect for Life

The pope termed this global culture of valuing money over people and putting everything up for sale as "a culture of waste":

> Unfortunately, in the countries of the so-called "developed" world, *an irrational consumerism* is spreading, a sort of "culture of waste," that is becoming a widespread lifestyle.
>
> JUBILEE OF THE AGRICULTURAL WORLD ADDRESS, VATICAN, NOVEMBER 11, 2000

The "culture of waste" is a poetic term but a disturbing reality. The physical effects of consumerism on the planet are obvious in how we are removing massive amounts of natural resources from the earth and quickly converting them into disposable packaging and products that within a year or less contribute to the growing mountains of waste in landfills, garbage slums, and toxic ocean gyres the size of Texas. We are converting our earth into waste, literally, daily. And that is not counting what we are blasting into the skies. We've got junk food, junk cars, junk bonds, junk mail...affluent houses are cluttered with junk while poor neighborhoods live with the landfills. *The Story of Stuff,* a short, quirky, animated Internet video on the typical life of a manufactured item, connects all the dots with humor (www.storyofstuff.org). But it's dark humor. It's enough to put one into despair. That's why Pope John Paul II also termed the culture of waste "the culture of death."

The pope warned about this spiritual disease of consumerism destroying our physical, social, and spiritual worlds:

> Equally worrying is the ecological question that accompanies the problem of consumerism and materialism, which is closely connected to it. In his desire to have and to enjoy rather than to be and to grow, man consumes the resources of the Earth and his own life in an excessive and disordered way.
>
> ENCYCLICAL, *CENTESIMUS ANNUS*, 1991

The desire to possess and consume spreads like a virus, infecting every aspect of life. Marketing and advertising feed like sharks on dissatisfaction and insecurity. As the pope described it, "the flood of publicity and the ceaseless and tempting offers of products" strive to convince each of us that we are not attractive enough, successful enough, or happy enough without these things. These kinds of messages in unlimited media doses, especially to youth, train generations to be dissatisfied, demanding, in debt, and dependent on goods for confidence and satisfaction. Money becomes not a means to a living, but an end in itself, so we can acquire more things, as well as a sense of being loved and accepted. In the process, John Paul II said we fall prey to "a frenzy for consumerism, so exhausting and joyless."

In their book *Your Money or Your Life,* Vicki Robin and Joe Dominguez muse on what has led people, especially Americans, into the trap of overconsumption:

> We have come to believe, deeply, that it is our *right* to consume. If we have the money, we can buy whatever we want, whether or not we need it, use it, or even enjoy it.... We buy everything from hope to happiness. We no longer live life. We consume it. Americans used to be "citizens." Now we are "consumers."

When we define ourselves by what we own or what we wear, our technological prowess or job status, we lose a sense of ourselves—our identity, capabilities, and creativity. Our sense that we are all children of God, with inherent value. Consumerism even affects marriages, where people look for the "best buy" in a spouse and want a "return" when the person doesn't live up to expectations. It affects parenting, where parents want to please children by purchasing all the right toys, lessons, and clothes so they fit in and succeed. It affects friends, when they are picked for how they look or what they own rather than their characters. We become slaves to exterior symbols of meaning, which leads us to an unending hunger for more. And the resulting waste piles up.

John Paul II explained:

> All of us experience firsthand the sad effects of this blind submission to pure consumerism: in the first place a crass materialism and at the same time a radical dissatisfaction because one quickly learns...that the more one possesses the more one wants, while deeper aspirations remain unsatisfied and perhaps even stifled.
>
> ENCYCLICAL, *SOLLICITUDO REI SOCIALIS*, DECEMBER 30, 1987

The pope counseled that we must get beyond the longing for more things and embrace the longing to become more as a person, more human, more free.

> It is a matter...not so much of "having more" as of "being more."...
> [Man] cannot become the slave of things, the slave of economic

systems, the slave of production, the slave of his own products. *A civilization purely materialistic in outline condemns a man to such slavery.*

ENCYCLICAL, *REDEMPTOR HOMINIS*, MARCH 4, 1979

Consumerism and Depression

In such a marketing-bound society, the entire panoply of seven deadly sins are glorified and used for entertainment: pride, greed, lust, wrath, sloth, envy, and gluttony. Humility, generosity, modesty, respect, patience, industry and perseverance, duty, honor, sacrifice, courage, kindness, responsibility, and limits are mocked. The humor of the day becomes put-downs, to see who will survive and who should be thrown off the island. Talk to any teenager who has access to Facebook, TV, music, or videos—or is attending middle school or high school—to find out how psychologically harsh the world is. Is it surprising that the rate of youth suicide is rising? This is the personal cost of the culture of waste: our children. John Paul II worried:

> *The spread of depressive states has become disturbing.* They reveal human, psychological and spiritual frailties which, at least in part, are induced by society. It is important to become aware of the effect on people of messages conveyed by the *media* which exalt consumerism, the immediate satisfaction of desires, and the race for ever greater material well-being. It is necessary to propose new ways so that each person may build his or her own personality by cultivating spiritual life, the foundation of a mature existence.
>
> 18TH ANNUAL CONFERENCE ON HEALTH AND PASTORAL CARE, NOVEMBER 14, 2003

The pope's heart went out especially to the world's youth:

> Dear young people, you are under threat from the bad use of advertising techniques, which plays upon the natural tendency to avoid effort and promises the immediate satisfaction of every desire. The consumerism that goes with it suggests that people should ever be seeking self-fulfillment especially in the enjoyment of material goods.

…Put yourselves on guard against the fraud of a world that wants to exploit or misdirect your energetic and powerful search for happiness and meaning. I say all this to you to express my great concern for you.

HOMILY AT MASS FOR THE FAMILIES, APOSTOLIC JOURNEY
TO NIGERIA, BENIN, GABON, AND EQUATORIAL GUINEA

As for parents, John Paul II knew they can try to fight against the tide, but without a cultural conversion, they can only do so much to protect their young. He was emphatic that we would not change our addictive, destructive behavior until we acknowledged it:

Modern society will find no solution to the ecological problem unless it takes a serious look at its lifestyle. In many parts of the world, society is given to instant gratification and consumerism while remaining indifferent to the damage that these cause.

MESSAGE FOR WORLD DAY OF PEACE, "PEACE WITH GOD THE
CREATOR, PEACE WITH ALL OF CREATION," JANUARY 1, 1990

Corporate Profit over Common Good:
The Economic Wasteland

The trouble is, consumerism makes the national and global banks and economies go round. Corporations seek unlimited "growth" in profits, and therefore must encourage consumerism, producing soon-to-be obsolete or disposable, overpackaged items to keep their markets and profits going. Building things to last is not valued or promoted. The corporate, national, political, and media leaders who make decisions regarding land, water, air, and workers are often far away and isolated from those affected by their poor choices—choices they justify under the rubric of "job creation" or "market needs" or "shareholder value."

Pope Francis attacked the immorality of such corporate decisions, banking systems, speculative financing, and investments that take place with little concern for environmental or human costs. Like Christ in the temple overthrowing the tables of the moneychangers, Francis railed against the idolatry of the marketplace, where unlimited freedom is offered to the markets and governments are not allowed to limit business and industry actions, where no one takes responsibility and where profit rules, rather than values—all leading to the trashing of people and the environment:

> The worship of the golden calf of old (cf. Ex. 32:15–34) has found a new and heartless image in the cult of money and the dictatorship of an economy which is faceless and lacking any truly humane goal.
>
> The worldwide financial and economic crisis seems to highlight their distortions and above all the gravely deficient human perspective, which reduces man to one of his needs alone, namely, consumption. Worse yet, human beings themselves are nowadays considered as consumer goods which can be used and thrown away. We have started a throw-away culture....
>
> Man is not in charge today, money is in charge, money rules. God our Father did not give the task of caring for the earth to money, but to us, to men and women: we have this task! Instead, men and women are sacrificed to the idols of profit and consumption: it is the "culture of waste."
>
> WORLD ENVIRONMENT DAY, JUNE 5, 2013

The new pope's rebuke sounds worthy of the saint who threw his clothes at his father in Assisi's plaza. When Pope Francis started cleaning house at the Vatican Bank, he called for complete global monetary and corporate reform. He demanded a reorientation of what we accept as the a societal norm and what provokes our outrage:

> If in so many parts of the world there are children who have nothing to eat, that's not news, it seems normal. It cannot be this way! Yet these things become the norm: that some homeless people die of cold on the streets is not news.
>
> In contrast, a ten-point drop on the stock markets of some cities is a tragedy. A person dying is not news, but if the stock markets drop ten points it is a tragedy! Thus people are disposed of, as if they were trash.
>
> IBID.

The Path to Transformation

What will it take for us to transform global, national, and media cultures of waste to ones of life? Do we need to be hit with a disaster in our own backyard before the ecological realities seem real? Pope John Paul II hoped and prayed that the call of faith and ethics would inspire us to act now in empathy with others and future generations—rather than waiting selfishly for the ecological damage to hit us first.

> The aspect of the conquest and exploitation of resources has become predominant and invasive, and today it has even reached the point of threatening the environment's hospitable aspect: *the environment as "resource" risks threatening the environment as "home."*
>
> ADDRESS TO UN CONFERENCE ON HEALTH AND ENVIRONMENT, MARCH 27, 1997

He advocated that corporate leaders, businesspeople, and government representatives once more return to the concept of business *as a service* and business leaders as people with a spiritual vocation. It is in people of this type—who balance their individual and family good with the community good and their ability to serve—that the pope invested some hope.

> All work is collaboration with God to perfect the nature he created, and it is a service to others. It is necessary, therefore, to work with love and out of love!
>
> ADDRESS TO CHRISTIAN WORKERS, DECEMBER 9, 1978

He explained that in each business and in each profession, there is the call to serve the individuals one encounters but there is also a larger calling, that of serving the common good and society as a whole:

> A second aspect of the businessperson's attitude of service is manifested in his or her responsibility toward society. It is well to recall that the progress of society ought to be oriented toward the common good of all citizens, that is, avoiding the temptation to turn the national community into something that is at the service of the particular

interests of the company.... What must characterize the businessperson is loyal openness to the just demands of the common good. That is a matter of wanting to make the company a factor of genuine growth in society.

ADDRESS TO BUSINESS LEADERS, DURANGO, MEXICO, MAY 9, 1990

Obviously, consumers, citizens, and stockholders must also make better ecological choices in business, politics, and their personal lives. John Paul II advocated that people need to become informed about issues *as a matter of conscience* and use their power in the marketplace for good:

> A great deal of educational and cultural work is urgently needed, including the education of consumers in the *responsible use of their power of choice,* the formation of a strong sense of responsibility among producers and among people in the mass media in particular, as well as the necessary intervention by public authorities.

ENCYCLICAL, *CENTESIMUS ANNUS,* 1991

Personal and Corporate Restraint: The Freeing Concept of "Enough"

John Paul II did *not* consider material goods as evils in and of themselves. In fact, he wanted more material goods for the many poor people of the world—more sharing from the affluent to those with less. But when an economy of wealth and waste prevails, the quality of life and the spiritual understanding of God decline:

> It is not wrong to want to live better; what is wrong is a style of life that is presumed to be better when it is directed towards "having" rather than "being," and which wants to have more, not in order to *be* more but in order to spend life in enjoyment *as an end in itself.*

ENCYCLICAL, *CENTESIMUS ANNUS,* 1991

Still, we all know that realigning our attitudes and lifestyles is harder when the media and economic cultures we live in are not seeing the connections or valuing things beyond profit. This means that in an age of increasing media, we need to define the "good life" differently and be taught the difference between needs and wants from an early age, along with "advertising survival skills" to avoid being duped by media messages of dissatisfaction. The pope called us to action:

> This tendency [towards the culture of waste] must be opposed. To teach a use of goods that never forgets either the limits of available resources or the poverty of so many human beings, and which consequently *tempers one's lifestyle with the duty of fraternal sharing,* is a true pedagogical challenge and a very far-sighted decision.
>
> JUBILEE OF THE AGRICULTURAL WORLD ADDRESS, VATICAN, NOVEMBER 11, 2000

Pope John Paul II advocated the power of joyful fasting—taking a rest from too much screen time, too much food, too many activities, too many products and things to care for, too many profits, too many debts—to appreciate what we have in one another and this world, and say thanks. This sacrifice helps build compassion and share resources with others in the world:

> Simplicity, moderation and discipline, as well as a spirit of sacrifice, must become a part of everyday life, lest all suffer the negative consequences of the careless habits of a few.
>
> MESSAGE FOR THE WORLD DAY OF PEACE, "PEACE WITH GOD THE CREATOR, PEACE WITH ALL OF CREATION," JANUARY 1, 1990

Mahatma Gandhi once said: "Live simply so others may simply live." Bill McKibben suggests that we need to bring back the concept of "enough" that guided earlier generations into financial stability and well-being. In his book *Enough: Staying Human in an Engineered Age* (2003), he writes, "We need to do an unlikely thing: We need to survey the world we now inhabit and proclaim it good. Good enough." And find joy in this.

In some cultures, simple living is called *mindfulness,* discerning what is most needed and being fully present in the moment, and the joy within it. It is difficult to overindulge or waste when you are savoring each moment with gratitude and peace. In the 1950s, Zen master Thich Nhat Hanh developed the term "engaged Buddhism," which maintains that "meditation, awareness of the moment, and compassionate action as a means of taking care of our lives and society are important elements of the spiritual path."

Being continually aware and grateful for all the beauty and gifts this earth has given us is key to fighting consumerism and addressing the ecological damage of a culture of waste. Also key is seeing the connection between one's choices and others' lives. On numerous occasions, especially on World Food Days, the pope explained how individual and communal restraint and sharing should work:

> It would be no small achievement for this World Food Day, if those who have an *abundance of material goods were to commit themselves to a reasonably austere lifestyle,* so that they can aid those who have nothing to eat. *If some would free themselves from excessively extravagant habits,* it would bring freedom to others, who could thus escape the devastating scourge of hunger and malnutrition.
>
> MESSAGE TO JACQUES DIOUF ON WORLD FOOD DAY, OCTOBER 4, 2000

Do Money and Things Buy Happiness?

While the culture of waste is leading to starvation and refugees in some countries, and obesity, heart disease, depression, and suicide in others, with destruction of the environment everywhere, it doesn't seem to be leading to greater happiness in the countries with "more."

International Gallup polls on happiness and well-being show consistently that many of the most market- and consumer-driven countries are *not* the happiest. In the latest global poll, conducted in 2010 via phone and email surveys of 136,000 people in 155 countries, Costa Rica ranked the highest in life satisfaction and quality of life in the Americas, as reported in *Forbes.* Canada followed, then Panama. The United States was not even in the running. In Europe, the order was Denmark, Finland, then Norway; in Africa, Malawi, Libya, Botswana; in Asia and the Pacific, New Zealand, Israel, and Australia.

The chief researcher at Gallup, Jim Harter, considered why certain countries rated higher than others: "One theory why is that the citizens in these countries tend to have their basic needs taken care of to a higher degree than in other countries."

So having *enough* to have physical basic needs met is one necessity for happiness. John Paul II stated that this is a responsibility of each nation, to help foster and ensure sufficient food and shelter, clean water and air, basic health care and an education, and dignified work to all its people, especially those with the least resources:

> When there is a question of defending the rights of individuals, the defenseless and the poor have a claim to special consideration. The richer class has many ways of shielding itself, and stands less in need of help from the State; whereas the mass of the poor have no resources of their own to fall back on, and must chiefly depend on the assistance of the State. It is for this reason that wage-earners, since they mostly belong to the latter class, should be specially cared for and protected by the Government.
>
> ENCYCLICAL, *CENTESIMUS ANNUS*, 1991

After basic needs are met, people can focus on nurturing and sustaining their social relationships and their environments. As reported in *Forbes*, Harter noticed that "Costa Rica ranks really high on social and psychological prosperity. It's probably things systemic to the society that make people over time develop better relationships, and put more value on relationships. Daily positive feelings rank really high there."

It is not surprising that many of the "happy" countries such as Costa Rica, New Zealand, and Denmark, which are known for their forests and wildlife and their movements toward sustainable economies, are caring for humans *and* their environments.

Transforming Trash to a Thriving Community of Life

As Paul Hawken noted in *Blessed Unrest,* inspiring, noble, exciting, and heroic tales of ecological, economic, and social transformations are all around us if we just have the eyes to see and give witness. Consider the story of Albina Ruiz, who was an engineering graduate student in Peru. The garbage produced in the city of Lima was overrunning the streets of a poor barrio, contaminating the groundwater and rivers, homes, and lives. One thousand tons of garbage a day mounted in steamy, smelly, toxic piles, and only half was being picked up by municipal workers.

So Ruiz set up a system to empower poor residents to become resources entrepreneurs within their community, getting paid to pick up other residents' trash and find uses for it. She educated neighborhood families about the dangers of pollution and waste, and encouraged them to pay $1.50 a month (the cost of one beer there) for the entrepreneurs' pickup services, covering the workers' salaries. Through this innovative system, she turned garbage into a source of work and income and dignity in the area, while cleaning up the environment and raising the quality of life for all.

Ciudad Saludable (Healthy City), as the project is called, encouraged individuals, especially women, to see the trash as a resource and take on tasks that had been neglected before. Some women expanded into running successful compost companies from the waste they collected, and the entire community is more knowledgeable, waste conscious, and caring about their environment.

The work of Ruiz and more than 1,500 entrepreneurs and staff is a practical, ecological example of a different view of waste and of poverty, with business and entrepreneurism seen as service. New forms of social entrepreneurism and natural capitalism are emerging that focus not just on profits but on relationships, seeking positive changes and the full development and participation of the laborers and community in good work. These new types of businesses seek to use all resources well and effectively, without degrading them, and whenever possible to restore them. John Paul II talked about this *relational* view in his "Gospel of Life" encyclical:

> Every man is his "brother's keeper," because God entrusts us to one another. And it is also in view of this entrusting that God gives everyone freedom, a freedom which possesses an inherently relational dimension. This is a great gift of the Creator, placed as it is at the service of the person and of his fulfillment through the gift of self and openness to others.
>
> ENCYCLICAL, *EVANGELIUM VITAE*, 1995

Just a sampling of other sustainable entrepreneurial projects around the world is exciting and inspiring. Riders for Health is an organization that uses a fleet of motorcycles to deliver medicine to rural areas in Africa. Proximity Designs works with subsistence farmers to find out their needs and then design low-cost solutions, such as a foot-powered irrigation pump and a solar light system, that can be sold at low prices through farm implement outlets. And in Bangladesh, a whole network of floating schools has been set up to accompany the movement of families down rivers.

From Waste to Wealth

As for the Ciudad Saludable model, it was so successful that Ruiz has replicated it in numerous other inner-city neighborhoods, and she was asked by the government of Peru to come up with a national waste strategy. The project has grown into an international nonprofit, Ciudad Saludable International, that has projects throughout Latin America and the world. It teaches seven "R" principles of sustainable development, which the global community can use as moral and practical guidelines:

> Reduce—avoid anything that generates an unnecessary waste
>
> Reject—do not use a product that pollutes the environment
>
> Reuse a product or material several times without treatment
>
> Repair—fix what can still be used, extending its useful life
>
> Recycle—take advantage of waste to produce new products
>
> Responsibility—act as citizens, businesses, professionals, leaders, and authorities, without causing damage to health and the environment, by changing habits and paradigms
>
> Reeducate—to provide a substantive change in environmental citizenship training from schools and communities.

These are some of the principles of John Paul II's "Gospel of Life" translated into practical action. They are also the principles of the Zero

Waste movement, which works not just with individuals and businesses but entire industries to save resources as well as money, producing income from industrial by-products. It urges the designing of all products, packaging, and processes to eliminate waste, and links the removal of waste with "clean production" to avoid producing toxins.

Future-oriented communities such as Northern Melbourne in Australia have designed landfills more like airports, with numerous specialized hubs that sort the waste materials much as the women in Ciudad Saludable do. Rather than "dumps," such operations are called "integrated resource management" or "eco-industrial" sites, which can derive income to run the system by repurposing some of the materials. To get manufacturers and corporations to take more responsibility, Germany has introduced the "Green Dot" (*Der Grüne Punkt*) system, which requires industry to collect and recycle 80% of the content of the packaging not being recycled by communities.

Holding industries responsible for the waste and packaging they produce has resulted in huge reductions in waste, and in the consequent burdens on communities, individuals, and the environment. States such as Minnesota have implemented "product stewardship" legislation, to hold manufacturers responsible for the whole life and disposal of a product, and other states have deposit laws and plastic bag bans. Because it saves money, between 70 and 80% of Fortune 500 corporations have launched some zero-waste initiatives, with companies like Anheuser-Busch, Epson, and Fetzer's Vineyards nearing or meeting their goals.

Passing On Stories of Solutions and Hope

Pope John Paul II encouraged the media to highlight positive environmental stories such as these. He urged everyone involved in the media to consider their ecological vocation and obligation to serve God and the greater good, to improve people's lives and uplift their minds, offering hope and positive solutions. They cannot be party to climate denials and the covering up of truths for profit, but must speak truth to power. The pope challenged media representatives to choose between promoting a culture of waste and death or one of life, putting away their sarcasm and negative attacks in order to provide a forum for civil discussion and education for the common good, objectively uncovering the causes of cold evil and its dangers around the world:

> Your work can be *a force for great good or great evil.* You yourselves know the dangers, as well as the splendid opportunities open to you. Communication products can be works of great beauty, revealing what is *noble and uplifting* in humanity and promoting what is just and fair and true.
>
> On the other hand, communications can appeal to and promote what is *debased* in people: dehumanized sex through pornography or through a casual attitude toward sex and human life; greed through materialism and consumerism or irresponsible individualism; anger and vengefulness through violence or self-righteousness.
>
> All the media of popular culture you represent can build or destroy, uplift or cast down. You have untold possibilities for good, ominous possibilities for destruction. It is the difference between death and life—the death or life of the spirit.
>
> ADDRESS TO 1,600 POLICY MAKERS IN TELEVISION, RADIO, MOTION PICTURES, AND THE PRINT MEDIA, REGISTRY HOTEL, LOS ANGELES, SEPTEMBER 15, 1987

The point is, as Pope John II reminded us, that when we see wasting things, species, and people as an option, we lose sight of God, and of God as working within all of creation. We lose a sense of caring for the poor and vulnerable. We lose a sense of justice. Bill Sheehan of the GrassRoots Recycling Network emphasizes this connection: "Zero Waste is linked to Environmental Justice because as long as officials are looking for places to get rid of the waste they will be looking for sites for mega-landfills or giant trash incinerators. All too often the sites for these undesirable activities end up in the poorest and most disenfranchised communities."

The Great Paradigm Shift

This shift of perspective, from accepting waste as the norm without any responsibility attached to it, to seeing it as a moral problem with opportunities to be addressed, mirrors the renewing of the human spirit within companies and communities, and the renewing of the environment as well. Suddenly we see the bigger picture, God's spiritual and physical energies and resources pulsing through all of creation. No one is left out. No energy or resources are wasted. All are respected and treasured. This shift restores our connections to God, to the plight of the suffering, to other species, and to all those who will come after us on this planet. Pope John Paul II called us to these daily transformations of thought and action:

> In order that the world may be habitable tomorrow and that everyone may find a place in it, I encourage public authorities and all men and women of good will to question themselves about their daily attitudes and decisions—which should not be dictated by an unlimited and unrestrained quest for material goods without regard for the surroundings in which we live and which should be capable of *responding to the basic needs of present and future generations.* This attention constitutes an essential dimension of *solidarity between generations.*
>
> ADDRESS TO PONTIFICAL ACADEMY OF SCIENCES, MARCH 12, 1999

I become here the voice of those who have no voice,
the voice of the innocent, who died because they lacked water and bread;
the voice of fathers and mothers who saw their children die
without understanding, or who will always see in their children
the after-effects of the hunger they have suffered.

ADDRESS TO THE PEOPLE OF OUAGADOUGOU, MAY 10, 1980,
REPEATED IN ADDRESS TO THE UN IN NAIROBI, AUGUST 18, 1985

The Ecological Connections to the Poor: Pope John Paul II's Defining Example

It is 1980, in Ouagadougou, Upper Volta (now Burkina Faso), one of the last countries on Pope John Paul II's first tour of Africa. He is visiting six of the nations of the western Sahel, the chest of Africa being slowly swallowed by the encroaching Sahara Desert. His response to the famine, suffering, and desertification he witnesses encapsulates many of the principles of his ecological vision.

Less than two years into his papacy, the new pope is vigorous, energetic, and dynamic. It's just weeks before his sixtieth birthday, and traditionalists don't know what to do with this man—he hikes and swims and travels more often than any pope ever, and to places they never considered visiting.

He knows suffering. Before he was even twenty, he endured the loss, one by one, of all his family. "I was not at my mother's death, I was not at my brother's death, I was not at my father's death," he remembered. "At twenty, I had already lost all the people I loved."

He survived enforced labor as a stone cutter in the Nazi work camps and being hit, nearly fatally, by a truck. He just barely escaped imprisonment as a hidden seminarian, and with martyred Jewish friends and Polish priests, he had no illusions about the horrors of occupation, genocide, and totalitarianism in his own beloved country. He witnessed the cost of religious and national intolerance and the truth of everyday evil socially fostered and taken to extremes, as well as the horrors of ideals gone bad, in the Soviet Communist occupation. Still, he believed in the love of God and the goodness of people.

The Faces of Ecological Disaster

Now he stands looking out at a crowd of people, the welcoming black faces who have experienced a new evil beyond his imagining—drought and famine and homelessness, where mothers walk miles in bare feet carrying their babies to try to find food, only to have them die in their arms of thirst or malnutrition along the way. Ten- and eleven-year-olds the size of toddlers for lack of food. Farmers who weep as the crops they try to grow shrivel from lack of rain. The trees are gone. The wildlife is gone. The grasses are dead. Their livestock are dying. The villagers who are still alive walk days to wait in a refugee camp, without tent or possessions, hoping for a cup of rice or gruel and a sip of water. They are homeless and hopeless, and the world at large does not hear their cries or see their pain. It does not see the connections between them.

John Paul II does. And his heart will never be the same. To the crowds and the press who are covering the events in the Sahel, he cries out to the human family for compassion, responsibility, and action:

> I cannot be silent when my brothers and sisters are threatened. I become here the voice of those who have no voice,...the voice of fathers and mothers who saw their children die without understanding, or who will always see in their children the after-effects of the hunger they have suffered; the voice of the generations to come, who must no longer live with this terrible threat weighing upon their lives. I launch an appeal to everyone! Let us not wait until the drought returns, terrible and devastating! Let us not wait for the sand to bring death again! Let us not allow the future of these peoples to remain jeopardized forever!
>
> ADDRESS AT OUAGADOUGOU AIRPORT, MAY 10, 1980.
> REPEATED IN UNITED NATIONS ADDRESS, NAIROBI, KENYA, AUGUST 18, 1985.

Not only does he engage with Catholics, he meets with African groups of Hindus, Muslims, and non-Catholic Christian denominations. They enchant him with their intelligence, warmth, caring, generosity, belief, and hope. He, himself, responds immediately with aid.

All around, he observes up close the realities of environmental degradation and climate change that few of us ever have or hope to. It is all the scientists have predicted it to be and more. It is poverty. It is hunger. It is suffering. It is ugliness. It is hell. And it is already here.

John Paul II's Foundation for the Sahel: His Model of Ecological Action

Pope John Paul couldn't get those people out of his mind or heart, but the world could. Sadly, the desertification and unrest kept increasing in those African countries. So in 1984 he started the nonprofit John Paul II Foundation for the Sahel, "to provide opportunities for sustainable growth in the ecologically distressed countries of the Sahel region" (PONTIFICAL COUNCIL COR UNUM): Burkina Faso, Cape Verde, Chad, Gambia, Guinea-Bissau, Mali, Mauritania, Niger, and Senegal.

Fearing northern imperialism in development and environmental projects, the pope set up the foundation so that it would be overseen by local bishops and community leaders. He particularly wanted the people to "feel like it was their own" and that they were partners and active leaders and participants in their own solutions. The Foundation is still going strong, with grants awarded to Sahel-based community organizations and municipality projects, NGOs, institutions, and government programs. Reforestation and training in sustainable agriculture, sharing of technical knowledge and engineering, and the digging of wells are prized projects, as is the education of youth and women.

A contemporary sample of the two hundred or so annual projects includes installing water pumps powered by solar panels; native tree plantings for biodiversity, fruits, oils, nuts, and fuel; radio programs on sustainable agriculture; rain collectors and irrigation canals; soil erosion prevention and re-grassing projects; and community schools, especially for girls.

The Foundation embodies so much of John Paul II's key ecological teachings: that all life issues are related and interconnected; that caring for the environment reduces the suffering of people and other species; that people must be coworkers with God in caring for the environment; that we need to get to work to restore what has been damaged; that developed countries must not dictate ecological solutions to others but instead share funding, expertise, and sustainable technology, letting the local people lead and participate in their own self-determination and that of their land and water; that women and children must be educated and involved; that alternative energies must be developed and used; that science and technology are useful when they are guided by ethics and an understanding of human dignity and needs; that all people must be included, regardless of religious or secular perspective, and no racism or intolerance is accepted; and last, but certainly not least, that "all work must be watered with prayer," as Pope Benedict said in 2012 of the continuing work of the Foundation. The Foundation continues to be supported strongly by Catholic congregations in Germany and the Italian Episcopal Conference.

Another defining quality of the projects is support of people of varied cultures, whether Christian or non-Christian: "A beautiful characteristic of the foundation is its openness to the different religions of the inhabitants, thus becoming an instrument of interreligious dialogue" (COR UNUM VATICAN WEBSITE).

Seeing the Personal Effects of Ecological Damage Around the World

The John Paul II Foundation for the Sahel was just the beginning of the pope's in-depth ecological work. Just as scientists measure the changes to the earth's systems and species through controlled studies, population surveys, and photographic images over time, John Paul knew it by people's lives. During his nearly twenty-seven years as pope, he made 104 pastoral trips, embracing diverse peoples in 129 different countries, often returning again and again. He tracked so many miles he could have gone around the globe twenty-eight times, or to the moon and back three times. Pope John Paul II viewed the environmental devastation through the lens of refugee camps, orphanages, hospitals, battlefields, and polluted rivers, and indigenous people losing their lands to logging, mining, and oil drilling. Blessed are the poor.

In the Sahel and other locales, it became very clear to the pope that the real price of degrading the earth is disease, despair, and death. The first people to feel the harm are always the poor. Where environmental degradation goes, refugees, wars, and unrest follow (and vice versa), in a moral downward spiral. Decisions that affect the land, water, and air are not just scientific and economic choices but deeply moral and spiritual ones. For when you don't care for the environment, you aren't caring for the people either. Today the Franciscan Action Network and Catholic Climate Change Covenant are trying to illuminate the human costs in their YouTube video *Who's Under Your Carbon Footprint?*

When John Paul returned to Africa in 1988, he told his diplomatic corps in Harare, Zimbabwe:

> The problem of hunger and the plight of refugees are directly related to the essential question of human rights. *All human beings have a fundamental right to what is necessary to sustain life.* To ignore this right in practice is to permit a radical discrimination. It is to condemn our brothers and sisters to extinction or to a subhuman existence.
>
> That is why the continuing state of famine in some regions, and the growing numbers of refugees in Africa and throughout the world, must weigh on the consciences of all who can and should work to remedy these situations. *Hunger in the world and the multifaceted problem of refugees are but two aspects—both very basic and important aspects—of the whole series of questions that must be faced* in order that the world find its proper balance in a new international order based on justice, solidarity and peace.
>
> SEPTEMBER 11, 1988

The Moral Nature of Ecology Confirmed

The visit to the Sahel was a defining moment of John Paul II's papacy. From that point onward, he taught even more passionately that all ecological issues have at their heart spiritual and moral ills, for all life issues are related. It's hardly surprising then that Pope John Paul II proclaimed to the world in 1990, "The ecological crisis is a moral crisis!"

What *is* surprising, though, is that he said it is also a spiritual and social *opportunity*. A global chance for hope. For the environmental crises are challenging us all to integrate not only new ways of thinking, seeing, praying, and acting, but also new ways of working, buying, and investing our money and time, rebuilding our relationships, calming our hearts, choosing our food, designing our energy sources, buildings, cities, and economies, re-greening our communities, restoring wild places.

Through his international work and travels, the pope heard about solutions and viewed some in action. He conferred with some of the most experienced and knowledgeable scientists on earth; the most innovative and hopeful activists, environmental organizations, and development specialists; the most service-oriented nuns, clergy, and lay volunteers; the most energetic and persevering community organizers, teachers, medical staff, agricultural workers, municipal designers, foresters, marine and wildlife conservation officers; the most ingenious and ethical entrepreneurs and business leaders; the most generous and dedicated philanthropists; the most profound and caring spiritual leaders from the world's religions. They were all making a difference. They just needed more workers in God's fields.

Pope John Paul II said to young farmers of Italy what he says to us:

> I hope that your discussions will bring about concrete ideas for
> the spread of an ecological culture. May the earth flourish again as a
> garden for all.

ADDRESS TO YOUTH MOVEMENT OF ITALIAN
NATIONAL FARMERS CONFERENCE, JANUARY 9, 1988

PART III

ECOLOGICAL VIOLENCE:
The Key Issues

*Blatant disrespect for the environment
will continue as long as the earth and its potential
are seen merely as objects of immediate use and consumption,
to be manipulated by an unbridled desire for profit.*

POST SYNOD APOSTOLIC EXHORTATION, *ECCLESIA IN ASIA*, 1998

We are now aware of the threats to entire regions
caused by inconsiderate exploitation
or uncontrolled pollution.
Protecting the world's forests,
stemming desertification and erosion,
avoiding the spread of toxic substances
harmful to man, animals and plants,
protecting the atmosphere,
all these can be accomplished only through
active and wise cooperation,
without borders or political power places.

ADDRESS TO GOVERNMENT OFFICIALS AND DIPLOMATS, ANTANANARIVO, MADAGASCAR, 1989

The Twelve Key Ecological Issues

The prophet Jeremiah complained to God:

> *How long must the land mourn,*
> *the grass of the whole countryside wither?*
> *Because of the wickedness of those who dwell in it*
> *beasts and birds disappear,*
> *for they say, "God does not care about our future."*

The Lord replied that He has not been blind, that He has seen how people

> *have trampled down my heritage;*
> *My delightful portion they have turned*
> *into a desert waste.*
> *They have made it a mournful waste,*
> *desolate before me.*
> *Desolate, the whole land,*

because no one takes it to heart....
They have sown wheat and reaped thorns,
they have tired themselves out for no purpose;
They are shamed by their harvest,
the burning anger of the Lord.

JEREMIAH 12:10-11, 13

You can't get more direct than this: God does not consider the land and the birds and the animals to be side issues. God is watching. God hears the cries of the land itself and all the species in it and all those who are hungry, thirsty, and in pain. This is the continuation of God's song of joy about the earth that he was singing in Job, but this song is also one of heartache. St. John Paul II, too, heard these cries around the world, and the voice of the Lord saying that this is all wrong, sinful, immoral, ugly, and sad. "The future starts today," John Paul told us, "not tomorrow."

> The human family is at a crossroads in its relationship to the natural environment. Not only is it necessary to increase efforts to educate the world's people in a keen awareness of solidarity and interdependence, it is also necessary to insist on their understanding of the interdependence of the various ecosystems and on the importance of the balance of these systems for human survival and well-being.
>
> ADDRESS TO THE PONTIFICAL ACADEMY OF SCIENCES,
STUDY SESSION ON CHEMICAL HAZARDS, 1993

During his papacy, the signs had already mounted to show that we were close to the tipping point and we had to act before the earth's systems had built up so much momentum in one direction that the most devastating domino effects could no longer be averted. He was urging the world to get to work:

> We are now aware of the threats to entire regions caused by inconsiderate exploitation or uncontrolled pollution.
>
> Protecting the world's forests; stemming desertification and erosion; avoiding the spread of toxic substances harmful to man, animals

and plants; protecting the atmosphere; all these can be accomplished only through active and wise cooperation, without borders or political power places.

ADDRESS TO GOVERNMENT OFFICIALS AND
DIPLOMATS, ANTANANARIVO, MADAGASCAR, 1989

Setting Visionary Goals of Solidarity That Stretch and Innovate

A full-scale, concerted global effort had to be launched—and is even more critical now—with the cooperation of international and national leaders, state and municipal governments, political parties, corporations and non-profits, all religious faiths of the world, families, and individuals. The United Nations has to set binding treaties and inspiring goals that pull the world together—because every party sees that their ultimate quality of life and survival depend upon it, as do their economies.

Pope John Paul II did not just lay out the moral, scriptural, and philo-sophical mandates of our ecological vocation; he also offered counsel on how to apply them to the key environmental issues of our time. He often addressed the United Nations, industry groups, world leaders, professional societies, and municipalities, as well as Catholic communities around the world, urging them all to new perspectives, higher standards, and hope. As all the issues are interconnected, they are not listed here in a specific order of priority, as if one could begin at the top and work down to the bottom. A holistic vision with a radical change of orientation needs to be worked on all at once, horizontally, promoting new goals, habits, policies, and products. If we do, each issue on its own will become more possible to address. Pope John Paul II explained:

> The question of the environment is closely related to other important social issues, insofar as the environment embraces all that surrounds us and all upon which human life depends.

MESSAGE TO PONTIFICAL COUNCIL FOR JUSTICE AND PEACE, NOVEMBER 4, 1999

Going deeper, John Paul II taught that our moral, spiritual, and physical health are all tied directly to the health of our natural surroundings, and to the suffering of people near and far away, whether we recognize it or not. As Martin Luther King, Jr., wrote in his "Letter from Birmingham Jail": "We are caught in an inescapable network of mutuality, tied in a single garment of destiny. Whatever affects one directly, affects all indirectly."

So this is a map of hope—Pope John Paul II's ecology of awareness and action on the twelve most pressing interrelated issues:

1. Greenhouse Effect, Climate Change, Alternative and Nuclear Energies
2. Deforestation, Desertification, and Reforestation
3. Oceans, Marine Life, and Those Who Live with Them
4. Scarcity of Clean, Fresh Water
5. Poverty, the Worst Pollution
6. Oppression and Exclusion of Women and Others
7. Population Growth, Limited Resources, and Responsible Parenthood
8. Farming and Conservation
9. Indigenous Peoples, Lands, and Environmental Racism
10. Loss of Biodiversity, Endangered Species, and Interdependence
11. Chemical and Industrial Pollution
12. War and Peace

In June 2002, in a Common Declaration on the Environment, Pope John Paul II and Ecumenical Patriarch Bartholomew I urged us to see our ecological crises as spiritual opportunities to know, love, and serve God more deeply, and thus have our lives transformed as joyous co-creators of good with God to renew the face of the earth:

> God has not abandoned the world. It is His will that His design and our hope for it will be realized through our cooperation in restoring its original harmony.
>
> In our own time we are witnessing a growth of an *ecological awareness* which needs to be encouraged, so that it will lead to practical programmes and initiatives.

...In this perspective, Christians and all other believers have a specific role to play in proclaiming moral values and in educating people in *ecological awareness*, which is none other than responsibility towards self, towards others, towards creation.

Then the two Christian spiritual leaders, one from the West and one from the East, laid out their religious and ethical guiding goals:

1. To think of the world's children when we reflect on and evaluate our options for action...
2. To be open to study the true values based on the natural law that sustains every human culture...
3. To use science and technology in a full and constructive way, while recognizing that the findings of science always have to be evaluated in the light of the centrality of the human person, of the common good, and of the inner purpose of creation...
4. To be humble regarding our idea of ownership and to be open to the demands of solidarity...
5. To acknowledge the diversity of situations and responsibilities in the work for a better world environment. We do not expect every person and every institution to assume the same burden. Everyone has a part to play, but for the demands of justice and charity to be respected the most affluent societies must carry the greater burden, and from them is demanded a sacrifice greater than can be offered by the poor. Religions, governments and institutions are faced by many different situations; but on the basis of the principle of subsidiarity all of them can take on some tasks, some part of the shared effort.
6. To promote a peaceful approach to disagreement about how to live on this earth, about how to share it and use it, about what to change and what to leave unchanged...

JOINT DECLARATION ON THE ENVIRONMENT WITH
ECUMENICAL PATRIARCH BARTHOLOMEW I, JUNE 10, 2002

The Power of Solidarność

Pope John Paul II believed deeply in this power of solidarity (*solidarność* in his native Polish), the strength of the community working together to build something new and worthwhile, for generations to come, a life's legacy. Not only does this solidarity strengthen our ability to slow climate change and other environmental disasters, but the very process of working through conflict to build solidarity enlarges and mature us as people:

> Solidarity—properly understood as a model of unity that can inspire the action of individuals, government authorities, international organizations and institutions, and all members of civil society—strives for the proper growth of peoples and nations, its objective being the good of each and every one.
>
> Solidarity, therefore, *by overcoming selfish attitudes regarding creation's natural order and its produce, safeguards the various ecosystems and their resources as well as the people who live there and their fundamental rights as individuals and community members.* If solidarity is firmly founded on this reference to the human person, with his or her nature and needs, it can draw together plans, norms, strategies and actions that are perfectly sustainable.
>
> MESSAGE TO JACQUES DIOUF ON WORLD FOOD DAY, OCTOBER 15, 2004

Obviously, people, nations, religions, and businesses don't just jump into relationships of solidarity, especially if it requires being less self-centered. It seems they only do it if they are desperate in fear, are inspired by noble values, see benefits to themselves, or are forced to do so—or a mix of all of the above. Pope John Paul II urged us to act in love, as that is the most sustainable motivator in the long run: love for God and Christ, love for nature, love for one another, love for ourselves and our families. John Paul II advocated that love is even more powerful than fear, a love that spawns responsibility, solidarity, and action. At the fiftieth anniversary celebration of the formation of the United Nations in 1995, the pope stated:

We must overcome our fear of the future. But we will not be able to overcome it completely unless we do so together. The "answer" to this fear is neither coercion nor repression, nor the imposition of one social "model" on the entire world.

The answer to the fear which darkens human existence at the end of the twentieth century is the common effort *to build the civilization of love*, founded on the universal values of peace, solidarity, justice, and liberty.

For John Paul II, and the Ecumenical Patriarch Bartholomew, the uniting motivation of love comes down to what all of humanity shares—love of our children. This love can help us create a new life-giving, powerful culture, courageous enough to face all our crises with hope and God's assistance, so we can be transformed in the process. They wrote:

A new approach and a new culture are needed, based on the centrality of the human person within creation and inspired by environmentally ethical behavior stemming from our triple relationship to God, to self and to creation....

[A]ware of the value of prayer, we must implore God the Creator to enlighten people everywhere regarding the duty to respect and carefully guard creation....

It is love for our children that will show us the path that we must follow into the future.

COMMON DECLARATION ON THE ENVIRONMENT, ROME-VENICE, JUNE 10, 2002

Despite an increasing sensitivity to ecology,
even the earth is suffering—
perhaps as never before in human history—
from climatic changes in the ecosystem,
thus raising questions about
the future of our planet.

SYNOD OF BISHOPS, *INSTRUMENTUM LABORIS*, 2001

1. Greenhouse Effect, Climate Change, Alternative and Nuclear Energies

In 2007, *National Geographic* proclaimed that we'd reached "The Great Thaw," and *TIME* announced: "GLOBAL WARMING. Be Worried. Be VERY worried." No longer a vague, distant concept, climate change is turning personal as the United States and other nations are feeling the tragic effects and economic costs of the intensifying storms and natural disasters climatologists warned about—hurricanes, tornadoes, droughts, wildfire, floods. Recognition of the escalating effects of climate chaos is rippling through markets and budgets. Military commanders all over the world are planning for how global warming will increase wars over water and land. The World Bank issued an urgent report in 2012: "Turn Down the Heat: Why a 4° Warmer World Must Be Avoided." Actuarial societies are advising large insurance companies not to insure new homes in certain coastal areas. Banks are not loaning money to European ski resorts at altitudes lower than five thousand feet. There is even a UBS Global Warming Index in the financial markets, allowing businesses to hedge weather futures.

In spring 2013, security expert Michael Klare, author of *Resource Wars,* described in *Salon* the social disturbances that will arise in this climate-shifting resource-scarce world:

Brace yourself. You may not be able to tell yet, but according to global experts and the U.S. intelligence community, the earth is already shifting under you. Whether you know it or not, you're on a new planet, a resource-shock world of a sort humanity has never before experienced.

Two nightmare scenarios—a global scarcity of vital resources *and* the onset of extreme climate change—are already beginning to converge and in the coming decades are likely to produce a tidal wave of unrest, rebellion, competition, and conflict. Just what this tsunami of disaster will look like may, as yet, be hard to discern, but experts warn of "water wars" over contested river systems, global food riots sparked by soaring prices for life's basics, mass migrations of climate refugees (with resulting anti-migrant violence), and the breakdown of social order or the collapse of states. At first, such mayhem is likely to arise largely in Africa, Central Asia, and other areas of the underdeveloped South, but in time *all* regions of the planet will be affected.

In the United States, even Newt Gingrich, once a vocal skeptic, stated that the scientific community has finally come to consensus that climate change is a reality and that humans *are* in part responsible and in peril, so we should provide global incentives for companies to change their carbon footprint. Others say we need carbon-cap goals and incentives with legislated limits and taxes—carrots and sticks—to make changes as swiftly as possible, as the fate of our irresponsibility is already bearing down upon us.

Drastic Change Needed Yesterday, Now We're Playing Catch-Up

What is missing is a unified international understanding that this is not only a question of absolute expedience, *but also one of morality, and faith*: a drastic change is needed in our ethical and religious perspectives to drive new national and international policies, as well as revised personal understandings and habits. Now. We have no time to lose. Polar explorers—including Will Steger, Annie Aggens, Paul Schurke, and Lonnie Dupre—have already witnessed the melting of huge ice sheets, glaciers, and tundra, and how devastating these changes are to regional peoples and wildlife. And these effects ripple out from the Arctic.

The seas are already warming and rising. Melting sea ice and tundra are starting to seep methane from buried mega-caches. Methane is a greenhouse gas twenty-five times more powerful than carbon. The methane's release will increase the heat in the atmosphere and oceans, melting the ice and frozen tundra faster, releasing more methane—in a fatal feedback loop.

Scientists know there is a tipping point, and that once we hit it, the chain reactions will be too intense to be stopped or slowed.

The Moral and Religious Responses

Back in 1983, Pope John Paul II commissioned the Pontifical Academy of Sciences to investigate "damage done to the environment by the increase of carbon dioxide and by the reduction of the ozone layer." Seven years later, the accumulated evidence was so substantial and disturbing, he proclaimed that the issues were now in crisis status.

> *The gradual depletion of the ozone layer and the related "greenhouse effect"* *has now reached crisis proportions* as a consequence of industrial growth, massive urban concentrations, and vastly increased energy needs. Industrial waste, the burning of fossil fuels, unrestricted deforestation, the use of certain types of herbicides, coolants, and propellants: *all of these are known to harm the atmosphere and environment.*
>
> The resulting meteorological and atmospheric changes range from health damage to the possible future submersion of low-lying lands. While in some cases, the damage already done may well be irreversible, in many other cases, it can still be halted.
>
> It is necessary, however, that *the entire human community—individuals, states, and international bodies—take seriously the responsibility that is theirs.*
>
> WORLD DAY OF PEACE 1990 MESSAGE, "PEACE WITH GOD THE CREATOR,
> PEACE WITH ALL CREATION," JANUARY 1, 1990

Some criticized that he was being led astray by "doomsday" scientists. Confident in the science, the pope was neither dissuaded nor derailed from his mission. He had already encountered people in his travels who were watching rising sea levels swallow their low-lying coastal lands, Southern hemisphere inhabitants whose eyes were damaged by light from the ozone hole, farmers suffering from famine because of increased drought and loss of trees. He anticipated the industry-led campaign to discount and dishonor the scientific findings, and he soundly reprimanded the deniers. John Paul avoided the political squabbles over human versus natural causes, acknowledging that climate changes were due to *both* nature and humans:

> Certainly there are elements linked to nature and its proper autonomy, against which it is difficult, if not impossible, to struggle.
>
> Nevertheless it is possible to say that *human behavior is sometimes the cause of serious ecological imbalance,* with particularly harmful and disastrous consequences in different countries and throughout the world. It suffices to mention armed conflict, the unbridled race for economic growth, inordinate use of resources, pollution of the atmosphere and water.

> ADDRESS TO THE PONTIFICAL ACADEMY OF SCIENCES, STUDY WEEK ON SCIENCE FOR SURVIVAL AND SUSTAINABLE DEVELOPMENT, MARCH 12, 1999

He called for *immediate* compassion and action: global cooperation for better policies and laws to slow climate change and pollution; the development of alternative energies, sustainable technologies and farming practices; sustainable building and public transportation designs; and massive reforestation. His faith and moral imperatives are even more urgent today.

The Need for Science and Global Solidarity of Action

In this "ecological emergency," Pope John Paul II praised scientists for their greenhouse and climate model projections, urging them to continue their awareness-building work and their journeys to search for more earth-harmonious paths:

Your efforts to work out reliable projections constitute a precious contribution to ensuring that individuals, especially those who have the responsibility of guiding the destiny of peoples, fully assume their responsibilities to future generations, removing the threats arising from negligence, gravely mistaken economic or political decisions, or lack of long-term planning. . . .

I therefore invite the scientific community to continue its research to better discern the causes of the imbalances linked to nature and to humanity, in order to anticipate them and to propose replacement solutions for situations that will become intolerable.

ADDRESS TO THE PONTIFICAL ACADEMY OF SCIENCES, STUDY WEEK ON
SCIENCE FOR SURVIVAL AND SUSTAINABLE DEVELOPMENT, MARCH 12, 1999

He called upon the global community—its religious leaders and communities, world leaders and governments, scientists and businesses, farmers, artists, civic planners, media representatives, and everyday citizens—to listen to the scientists and commit in solidarity to reaching carbon reduction goals and increasing ecological understanding and education:

It is becoming more apparent that an effective solution to the problems raised by the risk of atomic and atmospheric pollution and the deterioration of the general conditions of nature and human life can be provided only on the world level.

ADDRESS TO THE XXV SESSION OF THE CONFERENCE OF THE UN FOOD
AND AGRICULTURAL ORGANIZATION, NOVEMBER 16, 1989

Develop and Put Renewable Natural Energies to Work

Pope John Paul II used religious language to craft his message of urgency: "We are called to discover new sources of energy to replace those that are nonrenewable or insufficient" (ADDRESS TO THE PONTIFICAL ACADEMY OF SCIENCES, SEPTEMBER 26, 1986). In the area of innovation and new technologies, he acknowledged that the more developed countries clearly have the leading edge, and because of that edge, they have the moral responsibility to share their technologies and training with those less fortunate. John Paul mandated that those with power, finances, and technology must work in solidarity with local people, seeking solutions that fit area economies and situations, integrating regional experts, workers, and the poor into planning and economic benefits. The obvious choices for technology sharing are those that harness nature's *free* energy sources instead of those that work against nature and are so costly to process and use, and in addition cause expensive pollution problems in the extraction and burning processes.

For instance, the pope suggested, what could be more appropriate for African countries than solar energies, since communities in these areas are often isolated, with few trees and an abundance of sun? It is estimated that fourteen and half seconds' worth of the sun's energy directed at the earth is more than humanity uses daily. Shared solar grid stations, portable solar generators, solar lights, and solar ovens can be used in communities far from any traditional electricity grid. Areas with less continuous sunlight may have access to the power of the tides, winds, or the earth's own heat, as in Iceland. The manufacturing and distribution of new green technologies can offer boosts to struggling economies—especially when everyday people and small businesses can begin to produce energy for themselves, reducing their dependencies on others.

John Paul II meditated on the options at length:

> Through the centuries, humanity has developed the energy sources that it needed: from fire to nuclear energy. At the same time, industrialization led to ever-greater consumption, to the point that certain natural resources are nearly depleted.
>
> We must find new methods for using energy sources. Governments must develop a common energy policy so that energy produced in one region can be used in another. *The sun as a source of energy should be studied more carefully, as well as wind, sea, and geothermic sources. You have been studying biomass and photosynthesis.*
>
> Wood is one of the oldest sources and in developing countries will long remain the primary source *but it must not lead to deforestation resulting in ecological imbalances.* Botanists, ecologists, pedologists must work together.
>
> As regards waterfalls, coal, petroleum, and nuclear energy, the choice to use them is based on a series of factors. It is important to

consider regulations necessary to eliminate the risks of certain energy sources and to promote a sound ecological balance, and the protection of flora and fauna as well as natural beauty. The worker and surrounding populations *must not be put at risk.*

ADDRESS TO THE PONTIFICAL ACADEMY OF SCIENCES,
STUDY WEEK ON MANKIND AND ENERGY, NOVEMBER 14, 1980

Violent Fossil-Fuel Extraction Methods

Pope John Paul II made these statements of caution about the coal and oil industries even before the age of ever-more-violent extraction methods, such as mountaintop removal for coal, hydraulic fracturing (hydrofracking) for natural gas, tar sands oil, transnational oil pipelines, and increased ocean drilling. The *Exxon Valdez* accident did not happen until nine years after the pope's energy week statement, and he did not live to see the loss of life on the exploding British Petroleum oil rig that exploded in the Gulf of Mexico in 2010, and its two hundred million gallons of oil spilled.

However, during his lifetime, the Niger Delta in Nigeria had endured "the equivalent of the *Exxon Valdez* spill every year for 50 years by some estimates. The oil pours out nearly every week, and some swamps are long since lifeless," stated reporter Adam Nossiter in the *New York Times* at the time of the *Deepwater Horizon* spill. Nigerian government figures estimated then that an average of 11 million gallons were spilled per year, for a total of 546 million over the past 50 years. (At the time, 40% of the Nigerian oil at the time was going to the United States.) Nossiter explained:

> Perhaps no place on earth has been as battered by oil, experts say, leaving residents here astonished at the nonstop attention paid to the gusher half a world away in the Gulf of Mexico. It was only a few weeks ago, they say, that a burst pipe belonging to Royal Dutch Shell in the mangroves was finally shut after flowing for two months: now nothing living moves in a black-and-brown world once teeming with shrimp and crab.
>
> Not far away, there is still black crude on Gio Creek from an April spill, and just across the state line in Akwa Ibom the fishermen

curse their oil-blackened nets, doubly useless in a barren sea buffeted by a spill from an offshore ExxonMobil pipe in May that lasted for weeks. Local women who protested at one pipeline distribution site were beaten by soldiers guarding it. But because the people are poor and far from the media, no one was hearing about their decades of spills. "'Whatever we cry about,' says one villager, 'is not heard outside of here.'"

This is the kind of injustice—environmental racism and multinational corporate imperialism, in which local people and workers are harmed and have no voice—that the pope warned against. Communities near oil pipelines and pumping stations around the world have similar complaints about spills and leaks. In just the year and a half after the *Deepwater* explosion, there were six major pipeline spills, leaks, or eruptions, and six other major ocean spills (and these are only the big ones the press noted). Here's the toxic worst-hits list for 2010-2011, after the Gulf disaster:):

- ExxonMobil pipeline, Niger Delta, Nigeria, over 1 million gallons;
- Trans-Alaska pipeline, near Fort Greeley, 210,000 gallons;
- Chevron pipeline, Red Butte Creek, 16,800 to 21,000 gallons;
- Endbridge pipeline, creek flowing into Kalamazoo River, Michigan, 800,000 gallons;
- China National pipeline, Xing Harbor, China, 461,790 gallons spreading over 165 miles in the Yellow Sea;
- Rainbow pipeline, Peace River, Alberta, 1.17 million gallons;
- ExxonMobil pipeline, Yellowstone River, Montana, 42,000 gallons;
- Conoco Phillips/China National Offshore, Bohoi Bay, China, 3 leaks in a month and a half, over 6,300 gallons of oil;
- Royal Dutch Shell oil rig, North Sea, two leaks, 67,000 gallons;
- Chevron Brasil, Campos Basin, off Rio De Janeiro, 155,000 gallons;
- New Zealand cargo ship, 400 tons fuel oil;
- Royal Dutch Shell, Nigeria coast, 1.68 million gallons.

The leaks and spills keep coming, as they are considered just part of the everyday risks of the business—which is why environmentalists are so adamant about the unpublicized dangers and costs of pipelines and ocean drilling.

Fossil-Fuel Extraction Equals Water Waste and Pollution

Besides the ever-present threat of spills, ruptures, and explosions, all fossil fuel extraction methods utilize enormous amounts of water on an ongoing basis, diverting them from other uses and toxically polluting water sources and surrounding lands. In nearby areas, deaths from cancers and other diseases multiply. This is in addition to the negative effects on regional wildlife, beauty, and quality of life, as well as the substantial greenhouse gases emitted in the heavy-duty extraction processes and all those emitted in the burning of fuels so extracted.

Some people look at natural gas as the fossil-fuel bridge to a better tomorrow because it burns much cleaner than coal. This is true, but natural gas well digging and production give off carbon as well as cause methane leaks and water pollution, which override any gains in cleaner burning. Some of the very newest natural gas well drillings have lower methane seepage, but the established distribution sites, pipelines, and the old natural gas wells plus the new hydraulic fracturing methods have intense methane releases. And this does not address the toxic water issues. (Pope Francis himself has come out strongly against fracking because of its effect on water systems and the poor.)

John Paul II counseled in 1990 that nations should not just assume that new technological advances in extraction or technologies are safe:

> The State should also actively endeavor within its own territory to *prevent destruction of the atmosphere and biosphere,* by carefully monitoring, among other things, the impact of new technological or scientific advances.
>
> WORLD DAY OF PEACE 1990 MESSAGE, "PEACE WITH GOD THE CREATOR, PEACE WITH ALL CREATION," JANUARY 1, 1990

The Nuclear Option: The Toxins That Keep on Giving

To avoid greenhouse gases, many people are getting on the nuclear energy bandwagon as the "environmental" solution. But is it? First of all, fossil fuels are burned in the mining and processing of the uranium. Water sources are depleted and polluted by the cooling and storage of the nuclear rods and cores. Then there are the threats from terrorism, earthquakes, and accidents. These, of course, are all related to the extreme and long-lasting toxicity of the nuclear material itself. Radioactive materials degrade in half-lives, with their toxicity remaining for centuries and beyond. The nuclear option and its threats to humanity were questions that Pope John Paul II was forced to deeply ponder.

After the Chernobyl meltdown in 1986, clouds of nuclear material passed over and dropped radiation across parts of Russia, Ukraine, Belarus, and Europe. Estimates suggest that at least 5 million people were exposed, 3.4 million of them seriously, 1.2 million of whom were children. They suffered profusely from weakened immune systems and various illnesses. John Paul observed that Chernobyl "has become the symbol of the risks connected with the use of nuclear energy." On the fifteenth anniversary of the disaster, he met with children suffering from radiation sickness and told the world not to forget that nuclear waste and war are long-lastingly and devastatingly toxic:

> Recalling the tragic effects caused by the accident of the nuclear reactor in Chernobyl, let us think of the future generations that these children represent. We must prepare a future of peace, free of fear and similar threats.
>
> This is a task for everyone. For this to happen, there must be a combined technical, scientific and human effort to put every kind of energy at the service of peace, with respect for the needs of the human person and of nature. The future of the entire human race depends on this commitment.
>
> While we pray for the numerous victims of Chernobyl and for those who bear on their bodies the signs of such a dreadful disaster, let

us ask the Lord for light and support for those who, at various levels, are responsible for the destiny of mankind.

ON THE OCCASION OF THE FIFTEENTH ANNIVERSARY OF THE
CHERNOBYL NUCLEAR ACCIDENT, VATICAN CITY, APRIL 27, 2001

The pope was reminding us that if we choose to forget the past, we are destined to repeat it. Japan is reeling from this truth today. Since the 2011 nuclear meltdown at Fukushima after the tsunami, three molten nuclear cores and more than 400,000 tons of contaminated water have been stored on site, near the ocean. Thousands of tons of water are still needed each day to keep the reactors cool, adding to the tons of contaminated water stored. The tanks are linked with plastic piping, extremely vulnerable in earthquakes and storms. Also, due to the corroding power of the salt from the ocean, some of the water containers have sprung leaks, contaminating the grounds and washing out from the site. To stop some of the onsite groundwater contamination, the Tokyo Electric Power Company (TEPCO) is planning to create an ice barrier. They've built silt fences to keep the contaminated seawater in the harbor, but these already have burst holes in places and need constant reinforcement. An October 2013 typhoon washed seawater over cement protection dikes onto soil contaminated with 70 times the legal limit of nuclear matter. The storm waters picked up the toxic isotope Strontium-90 and flowed into other areas, contaminating them, then went back into the sea.

The site also contains 400 tons of spent fuel rods (plus over 1,300 fuel assemblies and 11,000 unused rods) that need to be removed carefully before another earthquake hits, because the storage facilities are unroofed and crumbling. As reported by *Japan Times* and Reuters, the spent fuel rods contain 14,000 times as much radiation as was exploded at Hiroshima. It will take a minimum of two years of risky removal procedures with cranes and then the transportation to another site (the Japanese ask: Where? And how safe will that be?), beginning in November 2013. If something breaks or explodes, it will cause an accident estimated to be 85 times worse than Chernobyl.

Many in Japan are worried—very worried. They do not trust TEPCO. And the world should be concerned too. It needs to send in experts and technology to assist Japan in preventing an accidental holocaust. In addition, the leaching water is already being carried by ocean currents and marine life to distant shores. Irradiated plastics and trash from the tsunami started reaching Hawaii early in 2013; tuna laced with traces of Fukushima-tainted radiation have been caught off California coasts. Scientists say that doses are low enough not to be health risks, but the reverberations have already begun.

These are the problems faced at just one nuclear site. As of January 2013, there were 437 nuclear power plants in the world, with 65 more under construction. In addition, uranium is as expensive to extract as nuclear power plants are to construct, maintain, and provide with security. No nuclear site on earth is guaranteed safety from earthquakes, hurricanes, tornadoes, accidents, or terrorism. Hardly an inexpensive, safe, efficient fuel source by any count.

Nuclear Imperialism and Injustice

A less dramatic but no less morally significant nuclear question is the transportation and long-term storage of the waste, which entail extensive public health, security, and maintenance costs. By continuing to invest in nuclear energy, we are *choosing to institutionalize denial and environmental racism*—because affluent cities, the large users of the energy, do not want the wastes in their backyards. Plans for storage invariably seek out less populated regions, most often the lands of indigenous peoples or the poor, who lack political power or a strong voice. Consider the Yucca Mountain storage proposal in the United States, or the islands in the Pacific as disposal sites. Pope John Paul II called the world to remember this ecological injustice and racism when he addressed the bishops of Asia in 1999: "The dumping of nuclear waste in the area constitutes an added danger to the health of the indigenous population."

Even prior to the eye-opening disaster at Chernobyl, the pope had addressed the director of the UN Atomic Energy Commission in September 1983 to proceed with caution. He spoke with great respect for the agency's work in using nuclear science for peaceful rather than violent ends, but he also subtly prompted those involved in nuclear energy to consider the implications of environmental racism and the ever-present underlying threat of nuclear war:

Whether the object is industrial projects for developing countries, nuclear reactors, or programmes for the improvement of society, the human person is the guiding criterion. No project, however technically perfect or industrially sound, is justifiable if it endangers the dignity and rights of the persons involved....

Such a reflection will not always be easy to make but it is necessary.... Thus you will measure the worth of a project by the impact it will have on cultural and other human values as well as on the economic and social well-being of a people or nation. In this way you place work in the wide and challenging context of the present and future good of the world....

Promotion of the common good in your work demands respect for the cultures of nations and peoples coupled to a sense of the solidarity of all peoples under the guidance of a common Father. The advancement of one nation can never be realized at the expense of another.

ADDRESS TO THE DIRECTOR GENERAL OF THE INERNATIONAL ATOMIC ENERGY COMMISSION AND OTHER UN AGENCIES AND PROGRAM DIRECTORS, SEPTEMBER 12, 1983

True to his roots, John Paul II counseled the Atomic Energy Commission and the other UN development agencies to look to St. Francis as a model of values in action:

Yes, the ideals of St. Francis of Assisi are a link spanning generations, uniting men and women of good will of all centuries in the quest for peace, whose spiritual goals are furthered by the honest efforts and hard and concerted work performed each day by the experts of so many fields and disciplines. It is in his spirit that I permit myself to speak of your contributions to the world, of what you are able to do for humanity, by working together, as brothers and sisters under the common Fatherhood of God:... Lord, make us effective servants of humanity, servants of life, servants of peace!

Technological Solutions That Pollute Can Also Cleanse

In counseling against polluting energy sources, it is not that Pope John Paul II did not appreciate how much they had done to improve standards of living, ease labors, and alleviate suffering. He did. And he wanted these types of benefits for the people of developing countries. However, the majority of present energy production and distribution methods, through corporations and electrical grids, have clearly not always been good for the poor, other species, ecological systems, and the future. Pope Benedict XVI drove this point home clearly in his encyclical *Caritas in Veritate* or "Love in Truth," when he emphasized how immoral the present fossil-fuel energy systems are:

> The fact that some States, power groups and companies hoard non-renewable energy resources represents a grave obstacle to development in poor countries....
>
> The international community has an urgent duty to find institutional means of regulating the exploitation of non-renewable resources, involving poor countries in the process, in order to plan together for the future....

He stated that human development must be based in love and a sense of brotherhood, of shared fate as humans and children of God. Poverty results from the world forgetting this truth, supporting economic and social structures that isolate people instead of bring them together. "The more we strive to secure a common good corresponding to the real needs of our neighbours," Pope Benedict said, "the more effectively we love them."

In this encyclical, Pope Benedict was building on Pope John Paul II's recurring challenge to us all to do better as a human family, embracing those with fewer resources into the planning and benefits of energy, economic, health, development, and environmental policies. As John Paul said:

> If humanity today *succeeds in combining the new scientific capacities with a strong ethical dimension,* it will certainly be able to promote the environment as a *home* and a *resource for all people,* and will be able to eliminate the causes of pollution and to guarantee adequate conditions

of hygiene and health for small groups as well as for vast human settlements.

Technology that pollutes can also cleanse.

ADDRESS TO UNITED NATIONS CONFERENCE ON ENVIRONMENT AND HEALTH, MARCH 24, 1997

That statement—*"Technology that pollutes can also cleanse"*—sums up his continuing optimism about human intelligence and potential and hope in God: we can't say, "No, it isn't possible to create better, more harmonious technologies because environmental standards will cost jobs." He believed that we have the ingenuity to do both, and more.

Confirming this, in January 2013, the nonprofit, nonpartisan research organization Media Matters for America issued "Myths and Facts on Solar Energy." The in-depth, up-to-date data demonstrated that solar energy is now economically viable and growing ever more so, with technologies at work in numerous countries and markets expanding exponentially. Many European countries are integrating alternative energies as part of their energy distribution policies and systems. Portable and small-grid solar and wind generators are proving to be lifesavers in places where infrastructures have been demolished by natural disasters or war, or in isolated villages that have never had access to an electrical grid. The sites are serving as test markets for technologies that can work in homes, businesses, and communities in developed countries as well.

In places where large wind farms or turbines are too loud or damaging to migrating birds, an updated version of the old-fashioned farm windmill, or a mini-turbine, is proving to be a cost-effective home or business generator, coupled with solar roof panels. Farm fields with grazing livestock can now do double duty as solar panel and wind farms. Thin, nano photovoltaic films make assembly of solar panels simpler and far more cost efficient and adaptable. They also promise to result in solar-power generating fabrics, car coatings, and roof sheetings.

The Need for Good Urban Planning for a Healthier Future

Pope John Paul II advised that we must not only use new renewable energies and technologies to slow climate change but also the power of restraint and design to conserve energy. In fall 2013, the Natural Resources Defense Council (NRDC) issued its first annual Energy and Environment Report, with the title "America's (Amazingly) Good Energy News." Its press release stated that "energy efficiency has contributed more to meeting America's needs than all other resources combined." 2012 energy consumption had dropped below 1999 levels even though population and economic productivity had risen. "The report stated that additional investments in efficiency could cut U.S. energy consumption by 23% by 2020, save customers nearly $700 billion, and create up to 900,000 direct jobs (plus countless more when consumers spend their savings elsewhere."

Like John Paul, Pope Benedict reminded the world of the importance of this kind of energy efficiency and conservation for the common good and alleviation of global poverty:

> The technologically advanced societies can and must lower their
> domestic energy consumption, either through an evolution in manu-
> facturing methods or through greater ecological sensitivity among
> their citizens. It should be added that at present it is possible to achieve
> improved energy efficiency while at the same time encouraging
> research into alternative forms of energy.
>
> ENCYCLICAL, *CARITAS IN VERITATE*, 2009

Cool Cities by Design

Since cities are where humans gather, it is where planning and building codes can make a huge difference in reducing energy costs and greenhouse gases—for instance, using white roofs in hot areas to reflect the sun's heat and black roofs in cool areas to absorb it. Pope John Paul II urged building and municipal designers to apply rigorous ecological ideals and standards, utilizing all the newest innovations and know-how, but also artistic guidance:

Good urban planning is an important part of environmental protection, and respect for the natural contours of the land is an indispensable prerequisite for ecologically sound development. The relationship between a good aesthetic education and the maintenance of a healthy environment cannot be overlooked.

WORLD DAY OF PEACE 1990 MESSAGE, "PEACE WITH GOD THE CREATOR,
PEACE WITH ALL CREATION," JANUARY 1, 1990

Effective public transportation, of course, is key to making a city livable. Who doesn't enjoy the ease of travel on the Metro in Paris, or the subway in London, Montreal, or New York, compared to the difficulties of travel in a metropolitan area like Los Angeles or St. Paul/Minneapolis, where there are no subways and insufficient (though improving) trains to outlying areas. Pockets of poverty fester when isolated from easy transportation to jobs, health-care clinics, banks, farmers' markets or supermarkets for a range of nutritious foods. Any young mother trying to haul her children to a grocery store or health center who has to endure hours on buses with numerous transfers can tell you how disabling it is not to have efficient public transportation. And any metropolitan area that depends mostly on cars can tell you the traumas of traffic and smog.

John Paul II, knowing this, challenged Rome and the cities of the world to invest in their people and infrastructure for moral reasons:

Public transport, in the current conditions of more intense movement of persons and often chaotic traffic, is destined to carry out a role of growing importance. There is a widespread need, from an ecological and human point of view, to guarantee greater "liveability" in our cities.

Our countrysides should not be further disturbed or polluted, and the human dimension of cities should be protected. And does all this not depend on the way that transport is organized? How important this is for Rome, given its joint role as capital of Italy and center of Christianity, does not need to be demonstrated.

JUBILEE ADDRESS TO ITALIAN GOVERNMENT WORKERS
AND OTHER GROUPS, NOVEMBER 25, 2000

And none other than Pope Francis has demonstrated the importance of public transport when he shocked the world with his fondness for riding the bus.

Making the Shift, the Leap of Faith to a New Future

In the end, if we truly want a new harmonious future with all of creation, we will have to risk the transition toward it. Pope John Paul II believed that human creativity can prevail if we put it to work in co-creatorship with God. Following his friend and mentor, Pope Benedict XVI had 2,7000 solar panels (donated by SolarWorld AG) installed on the Paul VI Audience Hall, saving 80 tons of oil and 225 tons of carbon dioxide emissions annually. (The idea originated in 2002 when John Paul II encouraged the founder of SolarWorld.)

In his 2009 *Caritas in Veritate*, Pope Benedict highlighted how the Church must "assert responsibility toward creation in the public sphere." He led by example. He gave major addresses on the environment both on his own and, as John Paul II had done, with Ecumenical Patriarch Bartholomew I, pushing for better international energy, financial, and environmental policies and more religious awareness of the need to care for creation and the world's poor.

The Vatican had once set its sights on being the world's first carbon neutral nation. The same year as the *Caritas in Veritate*, it announced an ambitious goal of developing Europe's largest solar-energy production facility, 740 acres of solar panels, in the community of Santa Maria di Galeria just north of Rome (site of the Vatican Radio transmitter) to provide sufficient energy for 40,000 homes. Little has been done since, it seems, so this plan is now left to Pope Francis to carry through.

Pope Benedict did have 580 square feet of solar panels put on the roof of his home in Regensburg, Germany (local plumbers donated the solar panels and young people at a trade school installed them), generating 5,800 kilowatt hours of energy annually, the equivalent of 11 barrels of oil. The money earned from selling the extra electricity to the German power grid is donated to education programs for disadvantaged youth. Pope Benedict also began using a donated hybrid electric popemobile. Pope Francis has followed his lead on this, using a rebuilt 1984 biofuel Renault and another all-electric Renault popemobile, as well as his old Taurus.

Catholic, Protestant, and other private and secular universities are transitioning to more sustainable energy systems, and some are starting to divest from the major fossil-fuel corporations. Hospitals are one of the largest commercial users of energy, and with the extensive Catholic health care system and ministries worldwide, the Church has enormous potential to lead the way in this transition to alternative energies, especially if the Catholic elementary education systems and parishes also did the same en masse. The Catholic Climate Covenant and the Franciscan Action Network, with their St. Francis Pledge and educational resources, are catalyzing the Roman Catholic movement. The Honorary Chairman of the Catholic Climate Covenant board is none other than the Most Reverend Bishop William Skylstad, Bishop Emeritus of Spokane and past President of the U.S. Conference of Catholic Bishops. He has long been working on environmental and social justice issues, calling on constituents to see the connections between these issues and their faith in Christ and our Creator.

Some Catholic parishes and Christian congregations have already gone "green" for God, moving away from fossil-fuels and disposable products because of their faith as well as because of the need to be frugal. Interfaith Power and Light is an American nonprofit organization that helps congregations transition to solar or other energies through its Cool Congregations program and its state affiliates.

The National Religious Coalition for Creation Care (NRCCC), working through its Green Churches program and alongside Interfaith Power and Light, has been encouraging similar energy transitions among other faith congregations. It has developed the "Carbon Confession," an examination of habits and conscience to help people of faith understand the carbon addiction built into our infrastructure, and make commitments to change. In the United States, the Environmental Protection Agency has been supporting all these efforts with guides for religious congregations wanting to make the shift to alternative energies.

Transitions can be an up-front burden for cash-strapped congregations (and households and businesses). In St. Paul, Minnesota, Father Kevin

McDonough has been leading St. Peter Claver Parish, a financially struggling but vibrant African American urban congregation, to live out its faith
and cut the costs of dealing with Minnesota winters. In a creative partnership, a group of parishioners have been working on forming a nonprofit
solar energy company to install solar panels on the church roof. If all goes
as planned, the church will then receive reduced electricity costs in a six–
year rent-to-own arrangement for the solar panels. Volunteers can even get
involved helping to install the panels. The company gets the roof real estate
and the rent to pay back the investment. It can sell the excess electricity back
to the grid and offer a working model of how other congregations can make
the switch. Everyone wins and God is served.

The Secular Shift for Finances and Ethics

Walmart and other large corporations are installing solar panels and utilizing alternative vehicle fuels, not for faith reasons clearly but for the bottom
line—and the long-term public good. They have set the goal of being supplied by 100% renewable energy in the future, and are already the largest
renewable energy generator in the United States, creating enough energy
to supply 78,000 homes. They have over 180 renewable energy generation
sites spread through seven countries, with more in construction, and they are
piloting mini-wind, large-scale wind, and solar/thermal programs in various
markets, including Canada, Chile, China, Mexico and the United States.
Since Walmart has the biggest truck fleet in the world, their transportation
goals will have a big impact, too. They are also engaged in a program with the
U.S. Fish and Wildlife Service called Acres for America, in which Walmart
helps create one acre of wildlife habitat for every acre of land it develops.

Small and mid-sized businesses are taking similar steps, as are municipalities. In the last two decades, over 1,100 communities in 43 countries have
committed to becoming Transition Towns or Cities, shifting away from fossil fuels toward alternative energies, local foods, community gardens, more
bicycling and walking, and improved public transportation systems. They
gain support, models, and resources through the networking of the Transition
movement. Transition Initiatives have begun in cities across the United States,

such as Ann Arbor and Asheville, Milwaukee and Missoula, Tampa and Tulsa, Salt Lake, Seattle and Santa Rosa. In the transition process, they have built deeper and broader community ties—and had some fun as well. They are showing that change is possible and beneficial on so many levels.

Worldwide Leadership for the Global Transition

Many transition efforts like these are working from the grass roots up. Yet for full-scale success, Pope John Paul II stated that world and national leaders will have to step up to the plate with honest, courageous, moral leadership and more ambitious international policies for the common good—and that these could bring indirect but important additional benefits:

> No plan or organization, however, will be able to effect the necessary changes *unless world leaders are truly convinced of the absolute need for this new solidarity, which is demanded of them by the ecological crisis and which is essential for peace.*
>
> This need presents new opportunities for strengthening cooperative and peaceful relations among States.
>
> WORLD DAY OF PEACE 1990 MESSAGE, "PEACE WITH GOD THE CREATOR, PEACE WITH ALL CREATION," JANUARY 1, 1990

He reiterated this theme in his 1991 encyclical on the new century, stating that each nation must take it upon itself to safeguard public goods like the atmosphere, oceans, air, fresh water, parks, and natural resources:

> It is the task of the State to provide for the defense and preservation of common goods such as the natural and human environments, which cannot be safeguarded simply by market forces. Just as in the time of primitive capitalism, the State had the duty of defending the basic rights of workers, so now, with the new capitalism, the State and all of society have the duty of defending those collective goods which, among others, constitute the essential framework for the legitimate pursuit of personal goals on the part of each individual....
>
> ENCYCLICAL, *CENTESIMUS ANNUS*, 1991

He was unambiguous: it is a moral duty of each state and all governmental representatives and agencies to work together to slow climate change and rebalance the atmospheric carbon equation.

The Kyoto Protocol

In 1997, the world's nations met at a UN conference in Kyoto, Japan to try to address climate change and fossil-fuel reduction. Since then, 192 countries have signed and ratified the UN Kyoto Protocol. The United States has remained one of the lone holdouts. Though it has been working on, and often meeting or even exceeding its targets, it has not wanted to commit to carbon reduction goals until developing countries, such as China and India, are required to do so. The developing nations, though, had lobbied for the freedom to develop with present fossil-fuel technologies to achieve lifestyle improvements enjoyed by the affluent developed nations, *before* they would be forced to reduce or cap their carbon emissions to below 1990 levels—so their reduction rates were voluntary and not enforceable by treaty or verified by objective reporting.

For nearly a decade previous, Pope John Paul II had been urging all nations to work together on reducing greenhouse gas emissions, requesting developed countries to lead the way in restraint and energy transformation, not waiting for other countries to commit to it first:

> *States must increasingly share responsibility, in complimentary ways, for the promotion of a natural and social environment that is both peaceful and healthy.* The newly industrialized States cannot, for example, be asked to apply restrictive environmental standards to their emerging industries unless the industrialized States first apply them within their own boundaries.
>
> At the same time, countries in the process of industrialization are not morally free to repeat the errors made in the past by others, and recklessly continue damaging the environment through industrial pollutants, radical deforestation, or unlimited exploitation of nonrenewable resources. In this context, there is urgent need to find a solution to the treatment and disposal of toxic wastes.

WORLD DAY OF PEACE 1990 MESSAGE, "PEACE WITH GOD THE CREATOR, PEACE WITH ALL CREATION," JANUARY 1, 1990

The pope's words were prescient. The path of merely voluntary caps on greenhouse gas emissions for developing countries led to a higher economic standard of living for many people in these nations but also the nations' ecological degradation. For example, the air, water, and soil pollution in China are so extreme that the toxins have resulted in mass public-health issues, sparking riots and unrest. A 2013 study showed that the air pollution from coal burning, road vehicles, and shipping traffic is cutting Chinese life expectancy by at least five years. That is why the new Chinese premier, Li Keqiang, pledged that cleaning up the country and integrating new ecological standards and technologies are key priorities for the nation.

In contrast, those countries, such as Germany, Italy, Norway, and Spain, that signed on to the Protocol in the 1990s have developed robust solar technology and wind industries. Iceland is already supplying 100% of its electricity needs from alternatives, relying largely on its huge geothermal source. In Germany, electricity generation is even shifting to the people themselves—about 51% of the nation's renewable energy in 2013 was produced by its citizens through solar and wind, with lower costs and greater personal control. China, sensing the change in economic breeze, has now become the biggest exporter of solar energy panels and wind turbines, fueling their economy with these new industries.

The United States, which led the pack of energy innovation decades ago and then lost political will, is now working to make a comeback. Wind supplies 5% of the nation's electricity generation, and this is growing rapidly, as is the percentage from solar. Jon Wellinghoff, chairman of the Federal Energy Regulatory Commission (FERC), said in August 2013 that the nation's solar industry is growing so fast that it looks to double itself every two years. With geothermal and hydroelectricity, 12% of U.S. power comes from alternative energies. The National Renewable Energy Lab found that the country can reach 80% of electrical production by renewable sources before 2050. But this may not be fast enough.

After the failure of full world assent to the Kyoto Protocol, Pope John Paul II summoned the international community at all levels (national leaders, legislators, business peoples, and nonprofits) to look beyond short-term economic values and choose sustainable, harmonious sources of power and industry, not just because it is the scientific necessity but because it is *the right thing to do, no matter one's religion, nation, or culture.* As he said to the Bishops of Asia:

> The protection of the environment is not only a *technical* question; it is also and above all an ethical issue. *All have a moral duty to care for the environment, not only for their own good but also for the good of future generations.*
>
> POST-SYNODAL APOSTOLIC EXHORTATION, *ECCLESIA IN ASIA*, 1999

International agreements and goals still lag behind potential and need. Because China had refused to submit to the Kyoto Protocol caps and reductions and the United States refused to sign, Canada pulled out in 2011, demanding a new agreement be made in which all nations have binding targets that are higher and more timely, to match the radical nature of the ecological emergency.

As we wait, the landscape continues to shift. Nitrogen trifluoride (NF_3), a "new" greenhouse gas, is not regulated in the present Kyoto Protocol. This gas traps 17,000 times more atmospheric heat than carbon dioxide, and it is presently emitted in the creation of LCD screens and some solar panels, which is obviously on the rise. There are processing alternatives, but without strict mention of NF_3 in a world agreement, its use will continue to increase.

The Doha Climate Conference resulted in plans to develop a full world agreement by 2015, with reduction goals set for 2020. Vatican City, considered an observer nation of the Kyoto Protocol, was not required to sign it and has not done so. Perhaps now is the time for Pope Francis and the Vatican to lead the world's nations in considering the moral and spiritual implications in this new treaty for more inclusive, stringent, and strident greenhouse gas cuts, taxes, trade incentives, debt forgiveness, and reforestation goals. (Perhaps after signing as a national leader, Pope Francis could be joined by other world religious leaders, such as the Dalai Lama, Patriarch Bartholomew I, and others, as advisory signatories.)

The Race to Carbon Balance and Moral Integrity

In addition to the Vatican, Costa Rica, New Zealand, Norway, Iceland, and the Maldives are in a race to become the first completely carbon-neutral, truly sustainable nations on earth. (British Columbia has stated that it has already achieved becoming the first carbon-neutral North American province or state.) Their efforts are renewing their economies and social structures in unexpected and exciting ways, establishing new methods in agriculture, industry, transportation, and building design that save money and the environment.

New Zealand's Prime Minister Helen Clark said: "We are neither an economic giant nor a global superpower. . . . If we want to influence other countries and the responses they take in coming years and decades, then we must take action ourselves. Taking action is not only the right thing to do, it is the smart thing to do" (AS REPORTED BY WORLDWATCH.ORG).

Pope John Paul II emphasized the interconnectedness of creation, and said that when we do the right thing and get into harmony with God and the earth, human economies, relationships, and quality of life are transformed:

> Reasonable and decent planning in the use of the planet's natural resources will greatly contribute to preserving nature, the human person, and his culture.
>
> ADDRESS TO PONTIFICAL ACADEMY OF SCIENCES AND THE PONTIFICAL COUNCIL FOR CULTURE SYMPOSIUM ON SCIENCE IN THE CONTEXT OF HUMAN CULTURE, OCTOBER 4, 1991

This year's theme for World Environment Day is
Environment and Peace and the motto:
"Let us plant the tree of peace."
Many trees will be planted to reforest the earth,
far too often stripped of its protective mantle.
This shows a love for nature and a commitment to protecting it.

GENERAL AUDIENCE, EVE OF WORLD ENVIRONMENT DAY, JUNE 4, 1986

2. Deforestation, Desertification, and Reforestation

It's a physical fact—what is put into the atmosphere is only one part of the greenhouse equation. The other part is what is absorbed out. The 2013 Report from the Intergovernmental Panel on Climate Change (IPCC) states, "Concentrations of CO_2, CH_4, and N_2O now substantially exceed the highest concentrations recorded in ice cores in the past 800,000 years." So cutting back to just pre-1990 levels is just not going to be ambitious enough. The IRCC report attributes the majority of precipitous rise in gases to "fossil fuel combustion and cement production" as well as "deforestation and other land uses." It also states that "most aspects of climate change will persist for centuries even if emissions of CO_2 are stopped." The atmosphere holds at least a century's worth of excess greenhouse gases that will continue to work their dark magic on the system unless they are absorbed.

But this is not a reason to despair. God evolved the planet to have its own built-in carbon-absorbers: plants, especially trees. It is no coincidence that at the heart of the Garden of Eden was the sacred Tree of Life, which was not to be touched. The answer to restoring balance is right in front of us—not only to stop fossil-fuel emissions but also to start major reforestation to absorb our past contributions, and re-green our planet.

Restoring the Tree of Life

A poet, Pope John Paul II evoked the tree as a symbol of how our personal and communal health is directly tied to the health of our natural surroundings and species, whether we recognize it or not:

> Unfortunately in our time, the tree is also an eloquent reflection of how man often treats his environment, God's creation. Dying trees are a silent warning that there are persons who obviously do not regard either life or creation as a gift, but only see what use can be made of them.
>
> It gradually becomes clear that wherever trees die, eventually man perishes, too.
>
> ADDRESS TO THE GERMAN DELEGATION ACCOMPANYING
> THE CHRISTMAS TREE TO ROME, DECEMBER 19, 1998

Certainly in the U.S. West and Southwest, where each year more and more intense and extensive wildfires rage, residents can feel the power of these words quite intimately. Though fire is a natural process and is needed for forest regrowth, too many fires and too-intense flames do the opposite. Devastating wildfires are sparked in the wake of clear-cut logging (rather than selective timbering), which leaves behind explosive tinder. In corporate timbering, if there is a subsequent replanting, the lands are often replanted with same-aged single species trees that burn at the same rates. This all spells trouble.

The increasingly hotter, drier summers and milder winters, combined with monoculture forests, have nurtured the expansion of diseases and insect eruptions, such as that of the mountain pine beetle. This small insect has turned Rocky Mountain vistas from evergreen to everbrown in a season, creating explosive fuel dumps. The types of wildfires that result in areas like this around the world are not completely "natural" disasters, they're ones we helped make. And suppressing forest fires in these conditions only makes them burn with more ferocity when they do erupt.

Without its forest canopy, the land has a harder time retaining moisture, and without reforestation and grassland restoration, it turns to unnatural desert. These are not gradual landscape shifts but radical ones that can be seen when comparing NASA photos taken over the decades. In Santa Fe and Albuquerque, seasonal dust storms have been multiplying and intensifying, with some fearing they will eventually drive residents away.

Pope John Paul II recognized the enormity of the wildfire and drought issues in his visits to the United States. In his 1999 apostolic letter to bishops and clergy of the Americas, he wrote of his concern for the immense amount of ecological destruction in many parts of America:

> It is enough to think of the uncontrolled emission of harmful gases or the dramatic phenomenon of forest fires, sometimes deliberately set by people driven by selfish interest. Devastations such as these could lead to the desertification of many parts of America, with the inevitable consequences of hunger and misery.
>
> POST-SYNOD APOSTOLIC EXHORTATION, *ECCLESIA IN AMERICA*, 1999

Because of deforestation, many regions in the world are also in danger of entering or reentering Dust Bowl eras. The tree serves both as a living metaphor and an ecological canary in the mineshaft. John Paul II stated:

> Since trees and plant life, as a whole, have an indispensable function in the balance of nature and are so necessary to life in all its stages, it is a matter of ever-greater importance for humankind that they be protected and respected.
>
> HOMILY ON THE FEAST DAY OF
> ST. JOHN GUALBERT, PATRON OF FORESTERS, VAL VISDENDE, ITALY, JULY 12, 1987

Forgiving Developing Nation Debts to Reduce Carbon in the Atmosphere

Sadly, many in-debt, developing nations promote the clearcutting or slashing and burning of intact forests for industry or agriculture so the governments will be able to pay their foreign debts—robbing Peter to pay Paul, so to speak. In many cases, the nations are former European or American colonies in Africa, Asia, South and Central America; so now, instead having colonial governments siphoning money to the colonizers, they have multinational corporations draining out profits from logging, mining, or plantations. Pope John Paul II points out the immorality of these scenarios, which simply tighten the circle of poverty in developing countries, leading to more suffering and loss of species:

> If an unjustified search for profit is sometimes responsible for deforestation of tropical ecosystems and the loss of their biodiversity, it is also true that a desperate fight against poverty threatens to deplete these important planet resources. Thus, while certain forms of industrial development have induced some countries to dramatically deplete the size of their tropical forests, foreign debt has forced other countries to administer unwisely their hardwood resources in the hope of reducing that debt.
>
> And likewise, the attempt to create lands for farming, pasture or grazing is sometimes an unfortunate proof of how inappropriate means can be used for good or even necessary aims. In this case, the solution of an urgent problem can create another equally serious one.

> ADDRESS TO PONTIFICAL ACADEMY OF SCIENCES AND ROYAL SWEDISH
> ACADEMY OF SCIENCES, STUDY WEEK ON "MAN AND HIS ENVIRONMENT,
> TROPICAL FORESTS AND CONSERVATION OF SPECIES, MAY 18,1990

This is one of the reasons why Popes John Paul II, Benedict XVI, and Francis have been so adamant about the necessity of the World Bank forgiving the debts of impoverished nations or rechanneling of loan payments in development and sustainability efforts. The National Religious Coalition for Creation Care (NRCCC) has been urging the World Bank to avoid offering loans for logging and deforestation, to instead invest in reforestation, and to disinvest from fossil fuel-based energy solutions—all for religious, moral, ecological, and economic reasons.

The NRCCC's sister organization, the Religious Campaign for Forest Conservation (RCFC)—an interfaith organization of Christian and Jewish leaders—drafted a religious "Vision for Forestry in the 21st Century" and offered a moral critique of World Bank policies regarding its forest loans to poor countries. With its challenge and the support of the Roman Catholics bishops of Mexico (who wrote a 2002 pastoral letter on forests), the RCFC won a $61 million grant and nonrepayable loan of $30 million to the government of Mexico for reforestation. The RCFC also set up an innovative pilot program in Guatemala that provided training in rainforest restoration to schoolteachers and students. They collected native seeds, raised the seedlings, planted them, and cared for the new trees. Needed school lunches were provided while the students learned about biology, ecology, and conservation.

The Benefits of Rainforest Biological Diversity for the World

The hallmark of rainforests is their diversity: it is estimated that the rainforests house 170,000 of the world's 250,000 known plant species. At least 80% of the diet of the developed world came from the rainforests. They provide important economic resources of wood, fruits, nuts, exotic flowers, rubber, spices, and so much more. Over 121 prescription drugs, including some of the most powerful cancer-fighting drugs, depend upon rainforest extracts as main ingredients.

Pope John Paul II highlighted this vast biological variety in his concerns about Amazonia: "This is one of the world's most precious natural regions because of its bio-diversity, which makes it vital for the environmental balance of the entire planet" (POST-SYNODAL APOSTOLIC EXHORTATION, *ECCLESIA IN AMERICA*, 1999).

Yet rainforests are being lost at a mind-boggling pace. The World Wildlife Fund reports that "some 46-58 million square miles of forest are lost each year—equivalent to 36 football fields every minute" to logging, ranching, palm oil and coffee plantations, among other agricultural enterprises (to which their soils may not even be suited). Some scientists estimate that we are losing 137 species a day, or 50,000 a year, due to these tropical forest losses—a prime example of the swiftness of ecological violence.

Because of the planetary significance of the tropical rainforests, Pope John Paul II called a pontifical scientific conference on them in coordination with the Royal Swedish Academy in 1990. He warned how we are indiscriminately fiddling with the very planetary systems upon which life has been built and depends, and stealing from present and future generations.

Unfortunately the rate at which these forests are being destroyed or altered is depleting their biodiversity so quickly that many species may never be cataloged or studied for their possible value to human beings.

Is it possible, then, that the indiscriminate destruction of tropical forests is going to prevent future generations from benefiting from the riches of these ecosystems in Asia, Africa, and Latin America? ... Should a lack of foresight continue to harm the dynamic processes of the earth, civilization, and human life itself?

ADDRESS TO PONTIFICAL ACADEMY OF SCIENCES AND ROYAL SWEDISH ACADEMY, STUDY WEEK ON MAN AND HIS ENVIRONMENT, TROPICAL FORESTS AND CONSERVATION OF SPECIES, MAY 18, 1990

Violence Against Forests Is Violence Against People

On a trip to Brazil in 2013, Pope Francis visited the shrine for the dark-skinned Our Lady of Aparecida, *Nossa Senhora Aparecida Conceição*. Here at this beloved shrine, he reaffirmed the need to preserve the Amazon for the poor people and species that call it home, and for the God who created them. In his radio address to the bishops of Brazil, he reminded them of the pledge they made to preserve the tropical forests at the shrine in 2007 in the *Aparecida Document*, which they signed with him and the bishops of Latin America and the Caribbean. Pope Francis called the Amazon a litmus test for the Church and for Brazil:

> I would like to invite everyone to reflect on what *Aparecida* said about the Amazon Basin, its forceful appeal for respect and protection of the entire creation which God has entrusted to man, not so that it be indiscriminately exploited, but rather made into a garden.

ARCHBISHOP'S HOUSE, RIO DE JANEIRO, JULY 28, 2013

As cardinal, he had been a prime mover in drafting the *Aparecida Document*, which stated unequivocally:

> Latin America has the most abundant aquifers on the planet, along with vast extensions of forest lands which are humanity's lungs. The world thus receives free of charge environmental services, benefits that

are not recognized economically. The region is affected by the warming of the earth and climate change caused primarily by the unsustainable way of life of industrialized countries....

The traditional communities have been practically excluded from decisions on the wealth of biodiversity and nature. Nature has been, and continues to be, assaulted. The land has been plundered. Water is being treated as though it were merchandise that could be traded by companies, and has been transformed into a good for which powerful nations compete. A major example of this situation is the Amazon....

As Popes Francis and John Paul II observed, as go the forests, so go their inhabitants, both wild and human. The timbering drives villagers, often indigenous people, from their traditional lands, stealing their rights and destroying their watersheds. Many of the large-scale logging operations are owned by international corporations such as Georgia-Pacific, Mitsubishi, Texaco, and Unocal. Violence against the land slides into violence against people as local civil-rights advocates, indigenous leaders, and everyday citizens protest. Many have been killed for standing up against corporations, government projects, or poaching rings in their areas. Father Nerilito Satur in the Philippines and Sister Dorothy Stang in Brazil are among the many who have become martyrs, assassinated for their faith and devotion to serving the poor, helping them protect their forests against logging. Sister Dorothy Stang was often photographed wearing a T-shirt with the slogan, "*A Morte da floresta é o fim da nossa vida*"—"The death of the forest is the end of our life."

Livelihood from a Live Forest

For local people, this statement about the intricate relationship between the rainforest and life is no exaggeration. Their lives and livelihoods have evolved together with the forest, and they have intimate knowledge of the practical and medicinal uses and ways of the plants and animals that outsiders do not. As the forests, languages, and people are lost, so is this knowledge.

Not surprisingly, leaving the forest intact and then selectively and sustainably harvesting species from within can raise the income and standard of living for all within rainforest communities. Leslie Taylor quotes in her book

The Healing Power of Rainforest Herbs evidence that in the Amazon, local people can earn five to ten times more by selective harvesting than by laboring at one of the big ranches, plantations, farms, or logging operations. An acre of living rainforest can be worth $2,400 a year in sustainably harvested fruits, nuts, medicinal extracts, insects, and other elements as compared to $60 an acre for cattle or $400 for one-time timber. But it is dependent upon setting up the right opportunities.

If these opportunities are developed, indigenous and local people can keep their lands while also having funds for necessities like education and health care. Nonprofit organizations such as Rainforest Action Network and CoolEarth support community-managed forests so that indigenous and local peoples can maintain both the forest and their lives. Pope John Paul II praised these types of development efforts and called upon forestry workers and scientists to continue educating people on the value of trees and their interdependencies with people. He pointed out that these forest crises are spiritual and communal opportunities:

> *An intense programme of information and education is needed.* In particular, your study and research can contribute to *fostering an enlightened moral commitment,* more urgent now than ever. . . .
>
> In this way, the present ecological crisis, especially grave in the case of the tropical forests, will become an occasion for a renewed consciousness of man's true place in this world and of his relationship to the environment.
>
> ADDRESS TO PONTIFICAL ACADEMY OF SCIENCES, STUDY WEEK ON MAN AND HIS ENVIRONMENT, TROPICAL FORESTS, AND CONSERVATION OF SPECIES, MAY 18, 1990

Forest Protection and Reforestation to Stop Carbon Overload

Pope John Paul II stood in awe of the mystery and power held within trees, and each encounter with a population starving because of drought-induced famine convinced him even more firmly of the importance of replanting trees in deforested areas and protecting all the forests we still have. In the John Paul II Foundation for the Sahel, he made community tree planting part of the key initiatives and projects:

Since the beginning of these painful events that make up the tragedy of the Sahel, plans have been drawn up for an attack on the drought and its causes and consequences, for bringing remedies to bear, such as irrigation, wells, *reforestation,* the construction of granaries, diversified crops and others. The needs are vast, however, if the advance of the desert is to be stopped and gradually pushed back and if every person is to have water and food and a future....

HOMILY AT OUAGADOUGOU, UPPER VOLTA, MAY 10, 1980

Success stories exist on a larger scale than his foundation, proving what can be done. Costa Rica lost almost 50% of its trees from 1950 to 1983 through logging, mining, coffee farming, and other plantations. But recognizing the devastation in the 1980s, it began reforesting. By 2010, it had replanted over one fourth of what had been lost. Unfortunately, many of the replanted forests are not as diverse, as nonnative, monoculture species were often used. However, the understanding of the great benefits of diversity has been changing replanting methods to include diverse native species. The nation's grand replanting has been accomplished through the collaborative work of the nation's Ministry of the Environment and its National Fund for Forestry Finance, with other partner programs.

The Green Belt Movement—
A Model in Community Conservation

One of the world's most extensive and effective community tree-planting projects was founded by the Kenyan Catholic ecologist Wangari Maathai. Her Green Belt Movement (GBM) trains and pays women and children small sums to help them raise, plant, and care for trees in their communities to save their watersheds, bind their soil so it doesn't wash away, and offer them shade, fruits, nuts, and fuel. "If you destroy the forest," taught Dr. Maathai, "then the river will stop flowing, the rains will become irregular, the crops will fail, and you will die of hunger and starvation."

Growing up in the steep foothills of Mount Kenya, Maathai was a member of the Kikuyu people, who believed in the sacredness of the fig tree, as it was "God's tree" and thus should never be cut. Yet Christian missionaries, fearing the people's religion, cut the trees, which led to ecological disasters. Maathai explained the functional importance of the fig tree—concepts the Kikuyu people understood in their cultural wisdom:

> [The fig trees] physically protected the land from sliding, because it's so steep. And because they have roots that go very deep and, as I say, because they are not cut, they last forever. They are able to go down into the underground rock...and they are able to bring some of the subterranean water system up nearer to the surface, and so they were responsible for many of the streams that dotted the landscape....they were part of the water system in the area, and so they served a very important purpose.
>
> INTERVIEW, *ON BEING WITH KRISTA TIPPETT*, WWW.ONBEING.ORG, 2006

Her Kikuyu elders taught her the sacred importance of trees and nature, while the Catholic school nuns demonstrated a devotion to Christ and service to others. In high school, Maathai joined the Legion of Mary, whose mission is "to serve God through serving fellow human beings." With her Catholic education, earned through scholarships, she discovered the field of ecology, where her faith and science were united. She became the first woman in Kenya to receive a PhD, and after much activism to improve the plight of the poor and the environment (for which she was persecuted, and even jailed), she was finally appointed Kenya's Minister of the Environment.

When Dr. Maathai explained her reasoning for setting up the Green Belt program to offer work to poor women, she echoed Pope John Paul II's counsel for engaging poor, local people as partners:

> If you want to save the environment, you should protect the people first, because human beings are part of biological diversity. And if we can't protect our own species, what's the point of protecting tree species?
>
> It sometimes looks as if poor people are destroying the environment. But they are so preoccupied with their survival that they are not concerned about the long-term damage they are doing to the environment simply to meet their most basic needs. So it is ironic that the

poor people who depend on the environment are also partly responsible for its destruction. That's why I insist that the living conditions of the poor must be improved if we really want to save our environment.

INTERVIEW BY ETHIRAJAN ANBARASAN, *UNESCO COURIER*, 1999

Since the founding of the GBM, ordinary citizens from poor villages in Kenya have planted more than 55 million trees, and the movement and its model have spread internationally. Their efforts have added to the biological diversity and overall ecological health of the areas in which they live. But equally important, and of more immediate relevance to these people is they have improved their own lives and those of their community beyond measure, providing food, beauty, clean air, and birds. They have improved their livelihoods through the sales of timber, firewood, fodder, and fruits; and enhanced the fertility and wildlife of their lands. They have cleaner watersheds with more abundant moisture; and more abundant building and fencing materials, among many other benefits.

Yet Dr. Maathai also found benefits of another nature:

The most notable achievement of the GBM in my view has been in raising environmental awareness among ordinary citizens, especially rural people. Different groups of people now realize that the environment is a concern for everybody and not simply a concern for the government. It is partly because of this awareness that we are now able to reach out to decision-makers in the government. Ordinary citizens are challenging them to protect the environment.

INTERVIEW BY ETHIRAJAN ANBARASAN, *UNESCO COURIER*, 1999

So not only is the Green Belt Movement planting trees, it is planting environmental awareness self confidence, and action while helping people more fully develop their potential and stronger community relationships— just as the pope had promised.

Greening the Ghetto Gives Back More Than It Takes

Majora Carter, founder of Sustainable South Bronx, pioneered ways to adapt Green Belt–like principles to plant elements of peace, prosperity, and hope in a rough urban area. In her 2009 article "Greening the Ghetto," Carter explained how she and her nonprofit used "the green economy as a social and economic solution to poverty":

> One of our first projects was the Bronx River. We knew that river restoration could start at any scale and grow. But we also saw that the labor to do the restoration was being imported—at a time when our neighborhood's unemployment rate was 25%. We put together a job training and placement system that eventually included riverbank and estuary stabilization, urban forestry, brownfield remediation and green-roof installation.
>
> Many of our participants were formerly incarcerated...those who have suffered the trauma of prison, poverty or combat do better when they work with living things and when they know that their work improves our collective society. That kind of work is extremely therapeutic. And they got paid!
>
> Since 2003, our Bronx Environmental Stewardship Training (BEST) program has maintained a job placement rate exceeding 80%, and 10% of our participants have gone on to college. It is one of the first and one of the most successful green jobs programs in the country. New York Mayor Michael Bloomberg recently followed our lead and unveiled a tree planting and maintenance job training program modeled on our efforts.
>
> POSTING BY MAJORA CARTER ON WWW.THEROOT.COM, APRIL 22, 2009

Carter preaches that what Sustainable South Bronx has done other urban areas can also do, "from planting gardens on our apartment rooftops to turning our dumpsites into parks,...city dwellers can save their communities and the world."

During the Great Depression, the United States pioneered the Civilian Conservation Corps (CCC) to put unemployed young men to work fighting the dust and repairing the stripped lands in a drought-punched nation. Many citizens have called for the return of this program as well as a national youth service corps. Any nation could form organizations similar to the CCC, employing their young adults to do tree planting and conservation work at this time of high levels of unemployment worldwide. This would jumpstart economies and reforest lands swiftly while offering the new workers pay, discipline, structure, training and education, and positive group identity and support.

Municipal Green Belt programs could also be created in developed countries as well as developing ones to get financially struggling women with children into gratifying work and conservation education, raising their self-esteem, means of livelihood, and community status. It establishes a family legacy of pride in work and a tradition of dignified community labor.

John Paul II affirmed the absolute importance and God-ordained dignity of this work:

> Conservation and management of woods, in whatever climate zone, are fundamental for the maintenance and restoration of the natural balances that are indispensable for life. This must be affirmed all the more today as we become aware how urgent it is that we decisively change all tendencies that lead to disturbing forms of pollution.
>
> Every single person is obliged to avoid decisions and actions that could damage the purity of the environment.
>
> HOMILY, ON THE FEAST DAY OF ST. JOHN GUALBERT, PATRON
> OF FORESTERS, VAL VISDENDE, ITALY, JULY 12, 1987

The Kingdom of God Is at Hand: Green Heaven

Dr. Maathai wrote *Replenishing the Earth: Spiritual Values for Healing Ourselves and the World,* and in it she noted how the many different religious cultures of the world hold the tree as sacred because of its power to support communities of life. She laid out a challenge to religious leaders:

> …when I am asked about heaven, I suggest that it might be green—a place of clean rivers with trees growing on the banks, fresh air, and all of nature's bounty on display.

And then I ask myself: Why can't we have such a life on this planet, right now? What is preventing us from cleaning our rivers, breathing fresh air, or growing food in abundance? Why do we have to wait until we get to heaven? The answer is almost always because we, ourselves, are doing things that are making that impossible: cutting down trees so that the rivers are silted with topsoil, producing green-house gases through burning fossil fuels, desertifying our pastures, and so on.

That said, the religious leaders have a special role because they are the ones who interpret the holy scriptures to the faithful and they ought to encourage the faithful to be custodians and caretakers of God's Creation.

EMAIL INTERVIEW WITH KATE MOOS, WWW.ONBEING.ORG, JANUARY 29, 2011

It is for just such reasons that Pope John Paul II stated that "the Creator entrusts the care of the earth" to us, and that in contemporary ecological activism and hands-on service, like tree planting, "the action of believers is more important than ever" (ECCLESIA IN AMERICA, 1999).

Be a Faith-Filled Forest Hummingbird

In 2011, Dr. Wangari Maathai died from complications of ovarian cancer. In her honor, a new initiative called the "I Am a Hummingbird" Campaign was launched by her friends and colleagues, with the goal of planting one billion trees in Kenya and around the world. The campaign was named after an old Kikuyu folktale she used to tell about a forest fire. When it broke out, none of the wild animals did anything but watch it burn because it seemed too big and ferocious to stop. Except for the hummingbird. She scooped up a drop of water in her small beak and threw it on the fire, zooming back and forth as fast as she could. When asked why she bothered, she said, "I am doing what I can." Through her example, the other animals followed her lead and, working together, defeated the fire.

Dr. Maathai explained:

In the course of history, there comes a time when humanity is called to shift to a new level of human consciousness, to reach higher moral ground, a time when we have to shed our fear and give hope to each other. That time is now.

Each of us can make a difference, and together can accomplish what might seem impossible.

THE GREEN BELT MOVEMENT, 2011 ANNUAL REPORT, WWW.GREENBELTMOVEMENT.ORG

At a crucial point in history, St. Francis was called by Christ on the cross: "Go, Francis, and repair my house, which as you see is falling into ruin." Now, God's house of earthly creation is also falling into ruin. Francis rebuilt the first little church in Assisi with his bare hands, all alone, but this example, his preaching, and his acts of kindness to all the needy and all of God's creatures gave others a model that started a movement. Tree by tree, prayer by prayer, we can do the same, on our own and with our families and friends, parishes and faith communities, cities and states, nations, and world. As we each do what we can, we inspire others to do the same, and with this comes a cultural shift.

To make this point, John Paul II also used a parable, only his was the story of the boy with the loaves and fishes whose small offerings multiplied to feed thousands:

> This…is the important task entrusted to you: like the boy in the Gospel, generously take the lead in a change that will mark your future…
>
> To do this, you must become conscious of what you possess, of your five loaves, your two fish, that is, of the resources of enthusiasm, courage and love that God has instilled in your hearts and in your hands, precious talents to be invested for others.

"LETTER TO THE YOUTH OF ROME," 12TH WORLD YOUTH DAY, ROME, SEPTEMBER 8, 1997

As a house is blessed, I have blessed the sea. . . .
The sea is not only a means of a living
but almost its own living reality adjoining the city. . . .
The sea is a heritage to be transmitted integrally to coming generations.

ADDRESS TO THE CEREMONY OF THE "MARRIAGE OF THE SEA," CERVIA, ITALY, MAY 11, 1986

3. Oceans, Marine Life, and Those Who Live with Them

In spring 2013, Ivan Macfadyen sailed the same ocean route he had taken ten years before, from Melbourne to Osaka via the California coast. But this time was different. "I've done a lot of miles on the ocean in my life and I'm used to seeing turtles, dolphins, sharks and big flurries of feeding birds. But this time, for 3000 nautical miles, there was nothing alive to be seen," he told *Newcastle Herald* reporter Greg Ray.

He heard silence where there had once been the calling of many birds, spouting of whales, splashing of dolphins. The only sounds were of discarded fishing buoys and trash hitting his boat. He and his mates could use their motor only during the day for fear of getting tangled in old nets. The paint of their craft changed color in the radiation-tainted water off Japan. Oil slicks, plastic, and garbage flowed with the currents instead of schools of fish, so they ate rice without fillets for dinner. They did, however, run into some fisherman who used deep nets, as Ray recounts:

> The speedboat came alongside and the Melanesian men aboard offered gifts of fruit and jars of jam and preserves. "And they gave us five big sugar-bags full of fish," [Macfadyen] said. "They were good, big fish, of all kinds. Some were fresh, but others had obviously been in the sun for a while.
>
> "We told them there was no way we could possibly use all those fish. There were just two of us, with no real place to store or keep

them. They just shrugged and told us to tip them overboard. That's what they would have done with them anyway, they said.

"They told us that his was just a small fraction of one day's by-catch. That they were only interested in tuna and to them, everything else was rubbish. It was all killed, all dumped. They just trawled that reef day and night and stripped it of every living thing." Macfadyen felt sick to his heart. That was one fishing boat among countless more working unseen beyond the horizon, many of them doing exactly the same thing.

No wonder the sea was dead. No wonder his baited lines caught nothing. There was nothing to catch....

The ocean is broken," he said, shaking his head in stunned disbelief.

GREG RAY, "THE OCEAN IS BROKEN," *NEWCASTLE HERALD*, OCTOBER 18, 2013

Similarly shocking snapshots of ocean distress are being recorded all over the world. In October 2013, all the sardine boats along the coast of British Columbia came back with empty hulls—not a single sardine. This is typically a $32 million dollar fishing industry, and of the fifty B.C. sardine licenses, about half are issued to First Nations fishermen. All lost not only this season's income, but the cost of the fuel, supplies, and labor. The hump-back whales, who usually follow the sardines and attract tourists, kept their distance from the usual coastal currents.

Sardine fishermen off South Africa experienced the same missing-fish woes. Many poor people around the world who depend upon fish for their suppers too often have empty stomachs. Beloved and crucial ocean species, including coral, sea turtles, dolphins, whales, sharks, manatee, tuna, cod, lobster, crabs, seabirds, and so many others are in critical danger.

The ocean is broken.

Pope John Paul II never lived near the ocean, but he knew its importance. What he said of the land at an address on the Monterey Peninsula in California applies to the ocean as well: it "will not continue to offer its harvest, except with *faithful stewardship*. We cannot say we love the land [or the ocean] and then take steps to destroy it for use by future generations" (SEPTEMBER 17, 1987).

In the Beginning Was the Sea

In the scriptural creation story, and in the evolutionary progress of the earth, the waters of the seas came before the land, and they were intimately connected to the waters in the sky:

> And God said, "Let there be a vault between the waters to separate water from water." So God made the vault and separated the water under the vault from the water above it. And it was so. God called the vault "sky." And there was evening, and there was morning—the second day.
>
> And God said, "Let the waters under the sky be gathered together into one place, and let the dry land appear." And it was so. God called the dry land Earth, and the waters that he gathered together were called Seas. And God saw that it was good.
>
> GENESIS 1:6–10

The ocean is the conjoined twin of the atmosphere *and* of the land. The blue of the sky reflects the glass of the sea, and the depths of the ocean mirror not only the physical shapes of the dry lands but their vast biological diversity, colors, and currents. Pope John Paul II reminded us of this in his letter to the bishops of Oceania: "The sea and the land, the water and the earth meet in endless ways, often striking the human eye with great splendour and beauty"(APOSTOLIC EXHORTATION, ECCLESIA IN OCEANIA, 2001).

Much of the world's seas remain mysterious and unexplored, with less than 5% studied—so in many ways we do not even know what we are destroying. What has been discovered is filled with wonders: fish as colorful as tropical birds. Undersea lakes and upside-down waterfalls. Kelp forests taller than our grandest redwoods. The world's longest mountain range, the Mid-Ocean ridge, crisscrossing the undersea globe for over 40,000 miles. Ocean caverns so deep that no light penetrates them, and we cannot yet plumb their depths. Yet marine life shines with bioluminescence in this darkness, lighting it like stars in the night sky or fireflies blinking on an evening breeze.

It is no wonder John Paul said, in awe of the sea and all of God's creation: "Ecology [of the sea] requires respect, admiration and affectionate attention, and it presupposes minds capable of contemplation, gratitude, and even poetry" (ADDRESS TO THE CEREMONY OF THE "MARRIAGE OF THE SEA," CERVIA, ITALY, MAY 11, 1986).

Five ocean bodies span the globe, but they are all really part of one large, interconnected planetary ocean. Covering over 70% of the earth's surface, the ocean encompasses nearly all the planet's water (97.3 %) and about half of its known species. It also produces half of the planet's oxygen.

The oceans, though, are also about people. More than one third of the world's population resides near the sea, and billions of people depend on fish and marine species to eat. One fifth of the protein in the global human diet is from marine species. Of the 925 million people suffering from chronic hunger and malnourishment, almost half (400 million) live in countries highly dependent on fish for their staple food source. So plummeting fish populations mean increased world hunger.

Averaged out, each person in the world consumes about thirty-five pounds of ocean food each year. Whatever toxins go into the ocean water go into the marine species we eat, ending up in the cells of our bodies. And the pollution, garbage, sewage, and nuclear waste we produce on land is washed or dumped into the oceans. So we are all linked to the oceans, for good and for ill.

Unfortunately, too few people are seeing the connections between the oceans and ourselves, or grasping the depths of the deadly challenges to the ocean systems and its species—and our own futures.

Out of Sight, Out of Mind, But Not Out of Trouble

The greatest threats to the oceans are the greatest threats on land:

- greenhouse gases causing acidifying, warming, desalinating and rising ocean waters;
- overharvesting, with harmful, wasteful, cruel, high-tech, and illegal fishing practices;
- garbage and waste proliferation, especially of plastics;
- chemical, mercury, oil, nuclear, and sewage pollution; and
- overdevelopment, destroying natural shore and tidal communities.

Even things we take for granted as nontoxic—the reverberating noises of ship traffic, sonar and radar signals, shore lights, and chemical smells—are interfering with the communication and mating of marine species, throwing them into additional decline.

In his 2001 *Ecclesia in Oceania*, Pope John Paul II wrote of the need for action to stop the damage to the Pacific Ocean, which represents what is happening in other areas:

> The natural resources of Oceania need to be protected against the harmful policies of some industrialized nations and increasingly powerful transnational corporations which can lead to deforestation, despoliation of the land, pollution of rivers by mining, over-fishing of profitable species, or fouling the fishing-grounds with industrial and nuclear waste. The dumping of nuclear waste in the area constitutes an added danger to the health of the indigenous population.
>
> APOSTOLIC EXHORTATION, *ECCLESIA IN OCEANIA*, 2001

During John Paul's papacy, the majority of ecological scientists and activists were so alarmed about greenhouse gases, the loss of the ozone layer, forests, and biodiversity on land, that the problems with the ocean systems, which are largely unseen, did not receive the same attention. John Paul II never summoned a Pontifical Academy of Sciences for a study week on ocean issues. But the time for action is upon us—from the Pontifical Academy of Sciences, the international community, world fisheries, and ordinary citizens. For as the oceans go, so shall we.

Changing the Ocean's Basic Chemistry

In many ways, the ocean serves as the moderating planetary thermostat, and over the recent century, it has also been our protective shield, buffering us from global warming's rapid, radical climate shifts, absorbing 30% of the excess atmospheric carbon and storing 80 to 90% of the extra heat. That is why the oceans themselves are warming.

Warmer ocean waters pulse more energy into the atmosphere, which affects weather. So as atmospheric and ocean temperatures are changing winds and currents, more intense hurricanes, tornadoes, blizzards, and monsoons are blasting the continents, or occurring in places they rarely have occurred. Superstorms like Typhoon Haiyan, Hurricanes Sandy, Katrina and Wilma, Super Cyclone Phailin that hit India, and Winter Cyclone Xythnia that pounded Europe are becoming commonplace. The changes in currents are also skewing marine migration paths and timing.

Like the forests and native grass communities on land, the ocean's kelp forests, sea plants, algae, and tidal grass communities absorb carbon and give back oxygen to the air and sea in a healthy exchange. But the saltwater also sponges up the atmosphere's carbon, and when the carbon mixes with the water, it turns into carbonic acid. So the unnatural excess of the airborne carbon is changing the chemistry of the ocean, slowly acidifying it. The ocean system is now 30% more acidic than before the industrial age and growing more so daily.

This acid then gradually dissolves the calcium in the bones and shells of marine species, slowing their growth. Imagine what that does to lobster, mussels, and coral reefs. Already oyster fishermen are finding far fewer seed oysters than in the past, as the shells are not being formed properly. Coral reefs are bleaching out and dying from the toxic chemicals drained, flushed, and washed into the seas. Now they are also being warmed and slowly steeped in carbonic acid.

Like floral gardens and tropical rainforests on land, coral reefs are among the most beautiful, extravagantly colorful ecosystems on earth. They are the nurseries of the seas, with over a million species beginning their lives or living within them. They support commercial, recreational, and subsistence fishing, especially in poor island nations. If the coral reefs die out, as they are in the process of doing, many struggling fish and other marine populations will collapse. The coral reefs also serve as tourist hot spots. In the Florida Keys alone, coral reefs generate over $1.6 billion in revenue annually.

The acidic water also disrupts the senses of marine animals, and the reproductive and protective strategies they've evolved. Clown fish, for example, are starting to swim toward predators instead of away from them because they can't "smell" them in acidic water. This is clearly not a healthy scenario.

Pope John Paul II's admonitions to shift from greenhouse-gas emitters to alternative energies are as urgent for ocean species as for those of land. In his apostolic letter to the bishops of Oceania in 2001, he called upon them and all the world to take responsibility for the Pacific Ocean and the other connective seas: "The continued health of this and other oceans is crucial for the welfare of peoples not only in Oceania but in every part of the world."

The Fate of the Fish and the Fishermen

According to Scripture (and science), before populating the earth, God populated the seas and the air.

> And God said, "Let the waters bring forth swarms of living creatures, and let birds fly above the earth and across the dome of the sky. So God created the great sea monsters and every living creature that moves, of every kind, with which the waters swarm, and every winged bird of every kind. And God saw that it was good. God blessed them, saying "Be fruitful and multiply, and fill the waters in the seas, and let birds multiply on the earth."
>
> GENESIS 1:20-22

However, in as much as God blessed and multiplied the oceans and its species, calling them all good, we as humans have been cursing the oceans into waste depositories and depleting species to extinction. Without coordinated international ecosystem management and changes in international fishing quotas and methods, many commercial marine species—such as bluefish tuna, shark, swordfish, cod, haddock, sea bass, hake, red snapper, orange roughy, grouper, grenadier, sturgeon, plaice, rockfish, and skate—will likely become extinct or critically endangered in the next 10 to 40 years. Some species are being caught or harvested at 2.5 times their reproduction rate. High tech radar, helicopters, and bottom-dragging trawling nets leave fish and marine wildlife few ways to escape. Some multinational fishing fleets have even been hauling in thousands of tons of catch beyond the market demand to deep-freeze and hoard for times of scarcity, or perhaps, extinction—when they will be able to corner the market.

Populations are being fished out and are collapsing in waters all over the world. The good news is, in places with effectively run and enforced fisheries, such as off the coasts of Alaska and New Zealand, thriving populations of key commercial species are being maintained.

Pope John Paul II spoke out about the enormous overfishing problems harming not only the wildlife but family and small-boat fishing operations. When speaking to Newfoundland fishermen, he said:

> With careful stewardship, the sea will continue to offer its harvest. However, during the last few years the means of processing and distributing food have become more technically sophisticated. The fishing industry has also been concentrated more and more in the hands of fewer and fewer people.
>
> Around the globe, more and more small or family fishing concerns lose their financial independence to the larger and capital-intensive enterprises. Large industrial fishing companies run the risk of losing contact with the fishermen and their personal and family needs. They are exposed to the temptation of responding only to the forces of the marketplace, thus lacking at times sufficient financial incentive to maintain production.
>
> Such a development would put the security and distribution of the world's food supply into ever-greater jeopardy, if food production becomes controlled by the profit motive of a few rather than by the needs of the many.
>
> ADDRESS TO THE FISHERMAN OF NEWFOUNDLAND, SEPTEMBER 12, 1984

Thirty years ago, the pope said it would take "courageous decisions" to rein in the corporate fishing fleet oligarchies that are driving family fishing boats off the ocean and decimating fish populations. He called for principles of ecology and justice to be applied nationally and internationally to the stewardship policies, limits, and licenses. He encouraged favoring small fishing operations over the multinationals, with the formation of fishing cooperatives for individual and family operators, and joint ownership that offers profit and decision sharing to the laborers in larger operations. Addressing marine fisheries, he called for immediate actions and a rearrangement of priorities to preserve the fish and ensure fishermen a voice in their own lives:

The Church is well aware of the difficulties and problems of the lives of those connected with the fishing industry, problems that are shared today by those throughout the world who earn their living from the sea....

The public authorities should favor forms of coresponsibility of those working in small-scale and large-scale fisheries, and the different forms of their solidarity in free associations. The active participation of all fishery workers in the decision-making that affects their lives and work should be encouraged.

One of your important tasks is to encourage appropriate use of available [marine] resources and develop new ones. Here, too, I would like to urge scientists to use all their talents and expertise. There must also be agreement on the criteria and methods to be applied to fishing in the context of world development....

ADDRESS TO THE WORLD CONFERENCE ON FISHERIES
MANAGEMENT AND DEVELOPMENT, JUNE 30, 1984

What have we accomplished in these thirty years?

Cruelty to Marine Species

The other moral and faith question that Pope John Paul II indirectly posed to the world is that of the suffering of fish and marine species from human treatment. Saints Francis and John Paul II stated that we must treat all species as our "brothers and sisters" deserving of care as God's creatures. *National Geographic* reporter Fen Montaigne wrote of practices on the ocean blue:

"Cruel" may seem a harsh indictment of the age-old profession of fishing—and certainly does not apply to all who practice the trade— but how else to portray the world's shark fishermen, who kill tens of millions of sharks a year, large numbers finned alive for shark-fin soup and allowed to sink to the bottom to die? How else to characterize the incalculable number of fish and other sea creatures scooped up in nets, allowed to suffocate, and dumped overboard as useless bycatch? Or the longline fisheries, whose miles and miles of baited hooks attract—and drown—creatures such as the loggerhead turtle and wandering albatross?

Do we countenance such loss because fish live in a world we cannot see? Would it be different if, as one conservationist fantasized, the fish wailed as we lifted them out of the water in nets? If the giant bluefin lived on land, its size, speed, and epic migrations would ensure its legendary status, with tourists flocking to photograph it in national parks. But because it lives in the sea, its majesty—comparable to that of a lion—lies largely beyond comprehension…

…all agree that the fundamental reform that must precede all others is…a change in people's minds. The world must begin viewing the creatures that inhabit the sea much as it looks at wildlife on land. Only when fish are seen as wild things deserving of protection, only when the Mediterranean bluefin is thought to be as magnificent as the Alaska grizzly or the African leopard, will depletion of the world's oceans come to an end.

HTTP://OCEAN.NATIONALGEOGRAPHIC.COM/OCEAN/GLOBAL-FISH-CRISIS-ARTICLE/

This radical change in orientation—looking beyond ourselves to what is under the waves not just with our eyes but with our hearts—is what Pope John Paul II said was one of the ways to "peace with God the Creator, peace with all of creation." He hoped that "St. Francis will help us to keep ever alive a sense of 'fraternity' with all those good a beautiful things the Almighty God has created. And may he remind us of our serious obligation to respect and watch over them with care…" (WORLD DAY OF PEACE MESSAGE, 1990).

Replacing Fish Food with Plastic

Perhaps even faster than we are removing fish and other species from the oceans, we are inserting plastic trash. We are multiplying the plastic garbage in the oceans so rapidly that the currents have swirled the waste into enormous floating gyres—two of them now larger than the state of Texas. They are miles deep, like upside-down mountains. Besides getting entangled in gyres, which become noxious funeral pyres, marine wildlife gobble the bags, toys, straws, packaging, and other waste, dying from the obstructions or from malnutrition. Rolled by the waves, the plastics break up and are massaged into colorful bite-sized beads. *There is now more plastic pellet "food" in the oceans than phytoplankton,* the building block of the marine food chain.

The plastics that are not eaten eventually degrade into a toxic sludge, eaten by the lower marine species, which are then eaten by those who are bigger, and then bigger, with the toxins traveling up through the food chain into the marine species we eat. Studies of the breastmilk of mothers who often eat sea food show traces of these toxins. *So unintentionally our infants and unborn are eating traces of our plastic trash.*

Mercury emitted by coal-burning utilities takes the same toxic travel route through the fish chains and later ends up our dinner plates and in breast milk. The rate of mercury poisoning for those eating a common diet of sea food is a growing public-health issue. What we put into the ocean system is what we get out, even if we don't intend to.

Aware of out many levels of connectedness to the oceans, in his address to Newfoundland fishermen John Paul II called for solidarity to care for both marine life and for the people dependent upon it:

> In a world of growing interdependence, the responsible stewardship of all the earth's resources, and especially food, requires long-range planning at the different levels of government, in cooperation with industry. It also requires effective international agreements on trade.
>
> PORT OF FLATROCK, NEWFOUNDLAND, SEPTEMBER 12, 1984

Restoring the Oceans

What can be done for the ocean's many ills? Ocean organizations—such as Ocean Conservancy, Oceana, and Blue Ocean Institute—are working to inform and engage the public, marine scientists and government agencies (such as NOAA), to repair and protect ocean system and species. They are advocating varied public policies that include:

- stringently reducing fossil fuel emissions and utilizing renewable energy technology;
- setting up blue carbon sequestration ocean areas;

- working out international ecosystem protection policies;
- establishing and investing in the enforcement of stringent agreed-upon international and national limits on commercial fishing;
- reducing fishing licenses and offering them first to subsistence, indigenous, individual, family, and smaller fishing operations;
- outlawing unsustainable and cruel fishing practices, especially those with great accidental deaths, or bycatch (such as longline fishing), and killing of predators;
- creating many more protected areas and marine reserve sanctuaries, especially in kelp beds and coral reefs;
- outlawing plastic nets, single use plastic bags, garbage dumping, nuclear and sewage drainage;
- monitoring cruise ship, barge, and shipping traffic emissions and outlawing their dumping;
- phasing out ocean oil drilling;
- transitioning out of chemical pesticides and fertilizer use inland;
- enforcing buffer vegetative zones on rivers and streams;
- restoring destroyed marine habitats;
- restricting sonar and radar use to necessities; and
- investing in more marine science.

Where does faith and action come in?

It is not enough, of course, for the responsibilities to be only on national and international policies and commercial fisherman to change. We all have to. The Marine Stewardship Council (MSC) helps consumers become informed about sustainable seafood and companies. That way those who purchase fish can reward those with good practices and avoid those with harmful ones. The MSC certifies fleets and products, crowning them with their label of approval to assure we can eat the seafood with a sense of peace. The Monterey Bay Aquarium has a similar Seafood Watch program, which recommends seafood that is fished or farmed according to sustainable practices. Communities of faith can help get the word out about these and other initiatives.

Ocean care doesn't end at the plate. People need to do what they can to slow climate change, reduce their use of plastics and packaging, avoid using chemicals on their lawns and fields, and much more. Our litter and trash must be picked up before it gets blown or washed out to sea. Parishes, congregations of faith, and families can get involved together in coastal and river cleanups, raising ocean awareness, and informing legislators.

Pope John Paul II encouraged all of us, but especially those in charge of oceans:

> The sea is a lovely image of this world in which our lives are lived. Humanity rides the billows of time to the shores of eternity....
>
> *I hope that all of you will live this relationship with natural resources fully respecting the marine environment* so that your work and a living will be safeguarded in peaceful coexistence for future generations, on the sea as on land, in nature and among human beings.
>
> ADDRESS TO GROUPS OF PILGRIMS FROM SPAIN AND ITALY,
> AND VARIOUS ASSOCIATIONS, DECEMBER 16, 2000

Toward a Faith-Filled Ethic of the Seas

To assist in this process, Dr. Carl Safina, founder of the Blue Ocean Institute, framed a secular ocean ethic to put sea issues into moral perspective as Aldo Leopold had done in his land ethic. In fall 2012, members of the National Religious Coalition for Creation Care (NRCCC) met in Hawaii to begin a similar process from a religious perspective, drafting an interfaith ocean ethic. They collected and read the latest scientific findings on the oceans and consulted with marine biologist Bill Thomas, director of the Pacific Division of the U.S. NOAA. He confirmed the need for the religious and moral ethic: "We've had the science information of what is happening and what needs to be done for years, but we haven't had the political will in the country or world to get it done. This is a moral issue, not a scientific one." And one of faith.

The committee strove for a succinct statement that would build on the wisdom of Aldo Leopold while encapsulating Judeo-Christian understandings alongside other spiritual guidance and common religious ideals in a faith-inspired ethic of the seas:

The whole cosmos, including the earth's lands and oceans, belongs to God, its ultimate Creator.

Thus, an act that affects the oceans is right if it tends to preserve, conserve, or repair the stability, integrity, beauty, and fruitfulness of the ocean's natural communities—of which we are an interactive part—and if it extends the Creator's respect, love, and kinship to all the living creatures therein.

Any act that tends otherwise is morally and religiously wrong.

And a status quo of harm is wrong as well. We are called as humans to love what God has given us, and work harmoniously with God for the healthy living of all species and systems, both now and to come, with special care for the poor, vulnerable and endangered. So we must begin our transitions, hard work, and sacrifices for better ocean and earth care now.

Together, and as individuals, we must ask for forgiveness, work to repair wrongs done and accept in humility that we will never live in perfect harmony, yet we must strive, and in the striving, and prayer, be transformed.

The Interfaith Ocean Ethic Campaign (www.oceanethicscampaign. org), run jointly by the NRCCC and the Franciscan Action Network, is striving to engage all churches, religious organizations, and believers in restoring and preserving the ocean systems and marine species. The organization hopes to catalyze the Roman Catholic Church along with other faith groups to make vital public statements nationally and internationally, and rally congregations and policymakers to act upon the "respect, admiration and affectionate attention" that Pope John Paul II said the seas deserve as part of God's creation.

David Krantz, a Jewish member of the Interfaith Ocean Ethic Campaign team, is a cofounder of the Green Zionist Alliance (GZA). Israel is home to four seas, all under threat. Water is declining in the Sea of Galilee and the Dead Sea, with advancing stretches of salty shore; pollution is attacking the Red Sea and Mediterranean, along with the Jordan River. The Green Zionist Alliance has prepared numerous action plans to reduce fossil fuels and re-green Israel. They have also developed a number of ethical and prayerful responses to articulate the Jewish community's relationship with the ocean, and help prompt and sustain caring actions and transformations: *Brit HaYam*: A Covenant with the Sea; *Birkat HaYam*: Blessing over the Sea; and *Tehillat HaYam*: A Psalm of the Sea:

TEHILLAT HAYAM: A PSALM OF THE SEA

Our fate rests with the sea:
With every breath
we breathe air from the sea.
Yet,

Listen to the sea:

It cries from overfishing—greed and gluttony.
It cries from pollution—avarice and wastefulness.
It cries from heat—carbon consumption and apathy.

May we return to days of old:
Let the sea and all within it thrive.

Befriend the plankton and the minnow,
The coral and the turtle,
The seal and the shark,
The dolphin and the whale;
Like us, they are part of Creation.

May the waves lap the shore
As a mother cradles her child.
Let the sea once again be filled with fish
As stars fill the sky.

Listen to the sea.

USED WITH PERMISSION OF DAVID KRANTZ

Pope John Paul II also, of course, advocated the power of prayer at all times and certainly for any journey of transformation. Devoted to the Blessed Mother Mary, the Star of the Sea, or *Stella Maris*, he prayed this for the people of Oceania. It can apply to us in our journey to care for the seas:

> O Stella Maris, *light of every ocean, and mistress of the deep,*
> *Guide the peoples of Oceania*
> *Across all dark and stormy seas,*
> *That they may reach the haven of peace and light,*
> *Prepared in Him who calmed the sea.*
> *Keep all your [our] children safe from harm for the waves are high*
> *And we are far from home....*
>
> *Bright Star of the Sea, guide us!*
> *Our Lady of Peace, pray for us!*

Human beings are thirsty for love and fraternal charity,
but there are entire peoples who go thirsty for life-giving water...
While the problem of progressive desertification
is also a real one in other parts of the world,
the suffering of the people of the Sahel impels me to speak of it here.

HOMILY AT OUAGADOUGOU, UPPER VOLTA, MAY 10, 1980

4. Scarcity of Clean, Fresh Water

In spring 2013, *Salon* ran the headline, "Will Water Supplies Provoke World War III?" Military specialists predict that they will. Already in the Middle East and other areas, there are conflicts over the distribution of water from shared rivers. Anyone who has been in the Rio Grande watershed for any amount of time can attest to the history of intense fights over water and the bloodshed that can sometimes result. With the ongoing droughts, the tension is only rising. It's happening all over the world.

Dr. Matthew Sleeth, author of *The Gospel According to the Earth* and founder of the organization Blessed Earth, tells in his book of an encounter with a graduate student from Korea who experienced culture shock from all the abundance she witnessed in the United States, from the home electricity down to the farm equipment. But it was our use of water that struck her most: "What is the strangest thing to me... is emptying my bladder in drinking water." That puts things in perspective pretty quickly.

The Sacredness of Water

As Pope John Paul II did with trees, he viewed water not only as a physical feature of nature, but also as a spiritual gift and symbol:

> Nor can we forget that water, a symbol used in the communal rites of many religions and cultures, signifies belonging and purification....

From its symbolic value springs an invitation to be fully aware of the importance of this precious commodity, and consequently to revise present patterns of behavior to guarantee, today and in the future, that *all people shall have access to the water indispensable for their needs,* and that productive activities, and agriculture in particular, shall enjoy adequate levels of this priceless resource....

MESSAGE TO JACQUES DIOUF ON 22ND WORLD FOOD DAY, OCTOBER 13, 2002

The pope reminded us that as children of the Creator, each of us has the God-given right to have what we need to grow. Therefore, all humans, no matter where they live or how poor they are, deserve the right to clean water—for themselves, for their children, and for generations to follow them. He stated explicitly in his 2003 World Day of Peace Message: "The right to safe drinking water is a universal and inalienable right."

This became a worldwide commitment in 2010 when the UN General Assembly passed the resolution affirming that access to clean water and sanitation *is* an inalienable human right and that all nations of the world have the responsibility to offer both to their citizens—and they must be safe, sufficient, acceptable, physically accessible, and affordable. But how?

The Scarcity of Clean Water

Even without droughts, clean water is scarce. Only about 2.5% of the world's water is without salt, and only about 1% is even accessible for drinking. Of that, only a small portion is drinkable without treatment. The unfortunate truth is that agricultural and lawn runoff, fracking, coal mining, oil pipelines, gasoline from barges, boats, and jet skis, chemical and medical waste from homes and factories, mercury, acid rain, and nuclear materials are toxifying the water in rivers, lakes, and underwater aquifers. The sewage washing into the water spreads diseases like cholera, typhoid, and dysentery. Pope John Paul II chastised nations that were not safeguarding their water resources and companies, such as transnational mining companies, whose practices polluted them.

Many people can't afford filtered or treated water—and sometimes even treatment cannot make polluted water safe. For instance, out of the hundred rivers in the Beijing area, only four are clean enough to even be considered for tap water, and those are being protected for the affluent. Yet people still have to have water, so they are forced to drink tainted water or buy it in bottles, which for so many is not a financial option.

The water advocacy organization Water.org reports that globally, 780 million people—two and half times the population of the United States—lack access to clean water, and 3.4 million people each year die of waterborne diseases. As droughts are increasing and deserts encroaching, consider the effort required to trek to a far away and unreliable water source, haul it back, and then perhaps boil it, if fuel is even available. Each day, women across the world spend over 200 million hours collecting and hauling water. This is becoming one of the biggest development challenges in every area of the world: how can we make clean water available to all?

In his Lenten Message of 1992, John Paul pleaded with Roman Catholics and other Christians, and all people of faith and good will, to seek solutions to the problems of desertification, pollution, and privatization of water.

> Remember those suffering from desertification and who lack that basic yet vital good—water. There is an expansion of the desert to lands that were once fertile. In many cases, people are the cause of the barrenness and the pollution of clean water.
>
> When people do not respect the goods of the earth, their actions result in poverty and death. Entire peoples have been reduced to destitution because of drought and water pollution. Uncontrolled development and the use of technologies that disrupt the balance of nature have caused grave disasters.
>
> I urge you to cooperate with those engaged in scientific analysis of all the questions for a solution to this problem.
>
> LENTEN MESSAGE, VATICAN CITY, 1993

The Privatization Push

Knowing the precious nature of water, then, the pope counseled that we need to use less of it and use it more wisely—putting a priority on citizen rather than corporate rights; protecting water sources through reforestation to hold moisture to the earth and the soil in place; stopping water pollution; and ensuring sanitation and affordable access to clean water for drinking and other *necessities,* as well as for wildlife. However, corporations such as Nestlé and Coca-Cola are buying up the rights to springs and clean water sources to supply the bottled water market. Businesses and investors are lobbying to privatize and run municipal water systems. In the past, the World Bank has supported initiatives of countries and municipalities to privatize their public water supplies in order for municipalities to have money invested into building or repairing the treatment, pumping, and piping infrastructure, but pressure is now being put on them to consider the moral and development ramifications of privatization.

As profit-seeking companies claim public water assets, clean water out of a bottle *or* the tap becomes a guarantee only for the affluent, while the local people of lesser means stand by, thirsty and desperate. It's an enormous moral crisis the world over.

In Rome, political fights over water broke out during the recent recession when the Italian government passed a bill requiring municipalities to privatize their water systems to save money. Citizens were enraged. So in 2011, a referendum to revoke the government's forced privatization was put on the ballot, and Roman Catholics and others rallied. Priests and nuns fasted and prayed at St. Peter's Square. Catholic lay people organized massive get-out-the-vote efforts, while bishops and Cardinal Peter Turkson reiterated to the government the Church's firm stances—based in justice, fraternity, and love—that water cannot be made into a marketplace commodity, but must be protected by the state as a public asset. At Assisi, Franciscans prayed to defend "sister water." The revocation referendum was soundly passed.

That fight, though, is going on in communities around the world. Citizens in Maine have been petitioning their governor not to contract their water rights away to Nestlé. For Popes John Paul II, Benedict XVI, and Francis, the only faith-driven, ethical answer is to guard water sources as gifts bestowed by the Creator for the common good, and pass them on in clean condition for future generations. Pope John Paul II made clear the ethical position that his successors built upon:

> Adequate levels of development in every geographical area will be legitimately and respectfully guaranteed *only if access to water is considered a right of individuals and peoples.*
>
> For this to happen, international policy must give fresh attention to the inestimable value of water resources, which are often *not* renewable and cannot become the patrimony of only a few since they are a common good of the whole of humanity. By their nature they "should be shared fairly by all mankind under the guidance of justice tempered by charity" (Second Vatican Council, *Gaudium et Spes*, 69).
>
> MESSAGE TO JACQUES DIOUF ON 22ND WORLD FOOD DAY, OCTOBER 13, 2002

John Paul knew the struggle would be long and hard. But Christ advised us to offer drink to those who are parched, thus we must act against the amoral drives of an impersonal, unlimited free market system:

> The international community and its agencies must intervene more effectively and visibly in this area. Such an intervention should be aimed at promoting greater cooperation in protecting water supplies from contamination and improper use, and from exploitation that aims only at profit and privilege.
>
> In these efforts, the primary objective of the international community must be the well-being of those people—men, women, children, families, communities—who live in the poorest parts of the world and therefore suffer most from any scarcity or misuse of water resources.
>
> MESSAGE TO JACQUES DIOUF ON 22ND WORLD FOOD DAY, OCTOBER 13, 2002

Bottles of Blue Gold, and Plastic Pollution

What happens if we do not?

In the documentary *Thirst*, filmmakers Alan Snitow and Deborah Kaufman reveal the "blue gold" dynamics of the new millennium marketplace—and the debilitating effects this is having on local economies and the poor worldwide, just as the pope warned.

When water is privatized and sold in plastic bottles there is another ripple effect of the disorder. The fossil fuels burned and oil-based plastics used in the making of the bottles is immense (47 million gallons of oil per year), and 80% of the bottles do not make it into recycling. Instead they end up in landfills or as land and ocean litter. It is estimated that a plastic bottle takes a thousand years to degrade fully. As bottled water is not regulated, there is no guarantee that the fluid is even as healthy as municipal tap water, and the quality of tap water gets neglected when there is a dependence on bottles. This is affluent convenience trumping the common good. Pope John Paul II was hopeful the tide was turning in favor of better water policies and protection.

> The growing awareness that water is a limited resource, but absolutely essential to food security, is leading many today to a change of attitude, a change which must be favoured for the sake of future generations....
> MESSAGE TO JACQUES DIOUF ON 22ND WORLD FOOD DAY, OCTOBER 13, 2002

Water Waste and Depletion in Aquifers

The deterioration and privatization of water from springs, rivers, lakes, and wetlands are not the only freshwater issues. In many places, water from the groundwater aquifers is being sucked up in enormous proportions for agriculture, industry, oil, coal, and gas extraction. In Chile, the Pinochet dictatorship offered corporations and private investors the rights to the nation's water for free for unlimited periods, leaving many communities, such as Mapuche indigenous villages, fighting for access to water as the aquifers beneath them are being siphoned down not only for bottled water but for mining and logging. For decades, the Columbian Fathers have been working alongside the native communities for justice, but it has been an uphill battle.

Tar sands oil extraction and hydraulic fracturing use and pollute exorbitant amounts of water from aquifers and watersheds. Fracking, as it is commonly referred to, works by blasting millions of gallons of water mixed with sand and chemicals at encased caches of natural gas and oil in order to release them. Not only does this waste clean water, but it injects toxic chemicals into rivers and aquifers, and also releases the gas or oil into ecosystems, making them flammable and toxic, thus endangering residents and wildlife in sur-rounding areas. On November 11, 2013, Pope Francis met at the Vatican with Argentine filmmaker Fernand "Pino" Solanas (*La Guerra del Fracking—The Fracking War*) and environmental activist Juan Pablo Olsson to discuss the severe damages that hydraulic fracturing does to water sources, and the suf-fering it cause the poor. Francis held up for the cameras t-shirts emblazoned with *No Al Fracking* (No to Fracking) and *El Agua Vale* + *Que El Oro* (Fresh Water Is Worth More Than Gold).

In addition, municipal water drawn from aquifers is often not only used for drinking water and home plumbing, but for lawns and commercial uses. Think of all the waste! This overuse is leading to severe depletions in many aquifers, especially when drought prevents rain from refilling them. Just con-sider the incongruence of Las Vegas's imposing casino fountains and Venetian canals in the midst of a semi-desert ecosystem where rivers and creeks often run dry and surrounding farm communities hope for rain and irrigation.

The United States has the biggest water footprint on the globe, and from 2000 to 2008, the nation's water use skyrocketed, depleting the nation's groundwater aquifers annually at an average volume of 20.2 million acres per foot deep. This is almost three times past rates, and we are not slowing down. We are using more water, and far faster than it can be replenished in a normal year of moisture, and the years have not been normal. Even in Minnesota, the "Land of Ten Thousand Lakes," communities have been complaining about their lakes draining as aquifer levels drop. Only coordi-nated ecosystem planning and shared sacrifice, based in a recognition of the preciousness of this gift for the common good, can avert the water losses, as Pope John Paul II pointed out:

> Because everyone must have access to uncontaminated water supplies, the international community is called to cooperate in protecting this precious resource from misuse and reckless exploitation.
>
> Without the inspiration of moral principles deeply rooted in human hearts and consciences, the agreement and harmony that should exist internationally in the preservation and use of this essential resource will be difficult to uphold and foster.
>
> MESSAGE ON THE 14TH WORLD FOOD DAY, OCTOBER 10, 1994

To Love and Restore

As a kayaker, the Pope John Paul II had a special affinity for rivers, not just as sources of drinking water but as places of scenic beauty, habitats for fish and wildlife, and symbols of the gifts of the Creator flowing every day. You can hear the deep love in his voice in the letter he wrote to the people of Poland on October 23, 1979, the day he was chosen to be pope. He expressed his sadness that he has to give up his beautiful homeland to stay in Rome:

> Dear fellow-countrymen, it is not easy to renounce returning to my country… To these mountains and valleys, to the lakes and rivers, to the people loved so much, to this royal city.

On October 9, 1988, during a trip to France, John Paul specifically took a trip down the polluted Rhine River— known as the "open sewer of Europe"—to call attention to its degraded state and support the plan to clean it up. Two years earlier, a fire at a Swiss chemical plant had spilled millions of gallons of toxic chemicals into the Rhine, killing fish and wildlife. This caused such an outcry that the international Rhine Action Programme (RAP) was established to coordinate and oversee efforts to drastically reduce toxic discharges, establish industry safety programs and municipal sewage treatment plants, build fish weirs and habitats, and replant native vegetation on the banks. Though salmon hadn't been caught in the Rhine since the 1930s, the RAP goal was to bring them back by 2000, because if sensitive salmon could survive, the river would be far healthier.

In Strasbourg, the pope boarded a Rhine River boat to tour the local area and bring media attention to the pollution and the restoration project. Nine nations are part of the Rhine River basin, and the long river passes through lands beautified by castles and vineyards, playing a key role in the history of Europe. Only in solidarity and collaboration could these nations restore and protect the river. This papal tour was not just about the Rhine, but about all the rivers of God's creation and about our human ability, and necessity, to work to restore and protect them.

The pope's faith in solidarity was well placed. By 1997, years ahead of schedule, salmon had returned. Industries had cut seepage into the river by more than 70% and agricultural runoff by 50% (though problems in that area still remain), and 63 of the native species had returned—some were even safe enough to eat! (Native sturgeons were still a holdout, but native mussels and other small creatures had come back but weren't yet edible.) Because of the weirs, migratory ocean-river species such as the salmon and sea trout were making their traditional spawning journeys through the river.

In 2002, the "Romantic Rhine"(the upper middle section of the river) was named a UNESCO World Heritage Site. The pope lived to hear of these triumphs of humanity and ecology he helped promote.

Rhine 2000 proved so successful that Rhine 2020 aims for the restoration of the full ecosystem—including groundwater and watersheds, flood protection, swimming spots, and all fauna in the river system (including the mussels) being safe to eat. Rhine 2020 is even utilizing nature-based treatments for purifying the drinking water.

A Stream Restored: A Story of Solidarity

The Rhine River was restored by coordinated initiatives overseen by an international commission, but this is not the only model. Reverend Owen D. Owens, chair of the Ecology and Racial Justice Program of the American Baptist Churches USA National Ministries, watched as the streams and rivers in his area outside of Philadelphia became sludged out with toxins, pushing out trout and driving people away. Being a man devoted to all of God's creation and an avid fly fisherman, Reverend Owens got to work in his own backyard.

He had cofounded the Valley Creek chapter of Trout Unlimited back in the 1970s, so he engaged his long-time fishing buddies in an effort to educate and inspire others to form the Valley Creek Restoration Partnership, thus bringing in people from all faiths and backgrounds. It was a decades-long project with many challenges along the way, but in solidarity, the community was able to restore Valley Creek to a Class A Wild Trout Stream. They even managed to convince the Walmart and Home Depot stores in the vicinity to plant trees and vegetative buffer zones between their parking lots and the riverbed. Reverend Owens captured this story in his book *Living Waters: How to Save Your Local Stream*: "We figured that if we tackled Valley

Creek it would have an impact throughout the state and across the country." His success showed how people of faith could build community relationships to restore habitats and wildlife.

This is just the type of service Pope John Paul II was advocating when he called all Roman Catholics, Christians, and people of good will from all faith and secular backgrounds to do what they can, wherever they are, to work together to restore our world. In 2004, the ailing pope wrote a message, entitled "Fraternity and Water," with the motto "Water, Source of Life," to encourage the bishops of Brazil in their Lenten campaign:

> Water is not an unlimited resource. Its...use calls for the collaboration of all people of good will with the proper governmental authorities to effectively protect the environment, considered as a gift of God.
>
> MESSAGE TO CARDINAL GERALDO MAJELLA AGNELO, MARCH 3, 2004,
> BRAZILIAN FRATERNITY CAMPAIGN, ZENITH.ORG

It is thus painful to note how many millions of
people are excluded from the table of creation.
For those people and for all the dispossessed of the world,
we must work hard and without delay so that they can
occupy their proper place at the table of creation.

LENTEN MESSAGE, VATICAN CITY, 1992

5. Poverty, the Worst Pollution

In 1972, when addressing the UN Conference on the Human Environment in Stockholm, Pope Paul VI stated, "Want, it has rightly been said, is the worst of pollutions."

In the 1970s and 1980s, the environmental movement was accused more than once of favoring other species over humans, and endangered animals over people in poverty. Since then, these accusations have fallen away as its successes have proven its worth and the movement has matured to be more inclusive in its mission, promoting sustainable development and ecological efforts among local people as a key tool to saving species, preserving biodiversity and wilderness, and preventing pollution. Concurrently, just as human development specialists have come to see that people's lives cannot be improved at the cost of their environment (and its wild species) or they will suffer even more in the long run.

In 1989, Pope John Paul II explained to the diplomatic corps of Holy See this interrelatedness of humanity with nature and how nations have to work together to support both at the same time:

> *A right to development and to the environment* are being spoken of today…this also shows humanity's growing awareness of interdependence with nature, whose resources—created for all but limited—must be protected especially through international cooperation.
>
> ADDRESS TO THE DIPLOMATIC CORPS, JANUARY 9, 1989

In his 1990 World Day of Peace Message he specifically noted that "The right to a safe environment is ever more insistently presented today as a right that must be included in an updated Charter of Human Rights" ("PEACE WITH GOD THE CREATOR, PEACE WITH ALL CREATION," JANUARY 1, 1990). The following year the UN Commission on Human Rights adopted Resolution 1991/44, which recognizes that "all individuals are entitled to live in an environment adequate for their health and well-being."

Despite this resolution, the pope continued to proselytize on the importance of linking development work and ecological restoration and protection, because all over the world, far too many poor people were living in areas denuded of trees, with toxin-laden soils, with chemicals and sewage in their rivers, or with intense urban smog. And far too few Catholics and other Christians, policymakers, government officials, and business leaders understood the connections between poverty and environmental degradation. People in affluent countries had to learn not feed like sharks on the meager resources of the poor nations or leave their pollution and waste with them. And they couldn't blame the poor for unwise ecological actions because people without resources or training sometimes inadvertently harm their own environments. John Paul II explained:

> The promotion of human dignity is linked to the right to a healthy environment, since this right highlights the dynamics of the relationship between the individual and society....
>
> The danger of serious damage to land and sea, and to the climate, flora and fauna, calls for a profound change in modern civilization's typical consumer life-style, particularly in the richer countries.
>
> Nor can we underestimate another risk, even if it is a less drastic one: people who live in poverty in rural areas can be driven by necessity to exploit beyond sustainable limits the little land which they have at their disposal. Special training aimed at teaching them how to harmonize the cultivation of the land with respect for the environment needs to be encouraged.

WORLD DAY OF PEACE 1999 MESSAGE, "RESPECT FOR HUMAN RIGHTS: THE SECRET OF TRUE PEACE," JANUARY 1, 1999

The God-made connections between human development and environment were the skeletal backbone of the theology of ecology, or care for creation, the pope was trying to explain; these are callings of faith and justice. Even in 2003 when he was aging and sick, he sent a message through the Holy See Delegation to the UN Economic Commission for Europe to outline three principles all development policies must aim to support:

- the *unique dignity* and the inalienable rights *of every human person,*
- the *unity of humankind,* constituted as a single family, within which all of us share in responsibility and solidarity for others,
- the *unity of all of creation,* which serves the needs of humankind, but which can never be considered just as the personal property of some, but is rather entrusted in stewardship to humankind for the good of its present and future generations...

...We must repeat that economic growth, in any part of the world, is not incompatible with the enhancement of an environment which is clean, healthy and is able to reflect the beauty given to it by its creator. Indeed it should be clear to all, including the business world, that globalisation will be sustainable in the long term, only in the manner in which it equitable integrates social and environmental concerns.

HOLY DELEGATION TO UN ECE FOR WORLD SUMMIT ON
SUSTAINABLE DEVELOPMENT, SEPTEMBER 24, 2001

The Poor As Creative Partners in a New Future

Pope John Paul II pushed the paradigm a step further. Those with few resources do not want to be powerless victims, he said, they want to be innovative, hardworking partners that care for their surroundings, not only for themselves, but for their children, grandchildren, and wild species. In his World Day of Peace Message of 2000, he proclaimed: *"Let us look at the poor not as a problem, but as people who can become the principal builders of a new and more human future for everyone."*

John Paul II expanded on this further in a speech to the UN Commission on Sustainable Development:

It is important to acknowledge that persons living in poverty must be considered as participating subjects. Individuals and peoples are not tools but protagonists of their future and agents of their own development. In their specific economic and political circumstances, they are to exercise the creativity that is characteristic of the human person and on which the wealth of a nation is dependent. *Sustainable development is aimed at inclusion.* It can only be attained *through responsible and equitable international cooperation, participation and partnership.*

HOLY SEE DELEGATION'S ADDRESS TO 11TH SESSION OF THE UNITED NATIONS COMMISSION ON SUSTAINABLE DEVELOPMENT. NEW YORK, APRIL 30, 2003

Consider the story of Kelvin Doe, a teen from Sierra Leone. He collected electronics from the landfills around his home, taking them apart to learn how they worked. His community needed lights, so he made his own battery out of garbage. His community needed a voice and music, so he created his own radio transmitter and began broadcasting. The International Development Innovation Network at MIT learned of this bright young man and brought him to the university to offer him greater opportunities.

In the inspiring YouTube video "15-Year-Old Kelvin Doe Wows MIT," Researcher Mark Feldmeier admits, "We kind of get trapped in our own little worlds. Just like Kelvin gets his world expanded by coming here, so am I getting my world expanded by working with him."

In this exchange, we can see the transformative nature of the sharing of expertise, values, and ideas. Development efforts that listen to the people's own ideas and concerns about what they need, and then share education, technology, and funds to help create sustainable futures have been proven to work for the long haul. Those that don't, end up bringing in technology from another country that works for a year or two until it breaks and there's no one to fix it and no spare parts. Then the technology becomes junk.

The pope explained how creative partnerships between local people and experts are more effective in solving the world's development *and* ecological crises:

These problems often require the expertise and assistance of scientists and technicians from industrialized countries. Yet the latter cannot solve them without the cooperation at every step of scientists and technicians from the countries being helped....

The training of local personnel makes it possible to adapt technology in a way that fully respects the cultural and social fabric of the local communities.... Only when this trained personnel finally exists locally can one speak of full collaboration between countries.

ADDRESS TO THE MEMBERS OF THE AGENCY OF THE
UNITED NATIONS, NAIROBI, KENYA, AUGUST 18, 1985

Forgive Us Our Debts, Lord, As We Forgive Others' Debts

However, this change of paradigm cannot, and will not, spread if those countries most in need (many of which are former colonies suffering from resource degradation and depletion) are bound in crippling debt. If they are sending large amounts of their general funds out of the country in loan payments to institutions of affluence, how will they support all the projects that are expected of them to develop and lift their citizens out of poverty and dependence?

John Paul II repeated the words of his mentor and predecessor Paul VI about poverty being the worst of pollutions and built upon them, describing the "geography of hunger," wherein rich nations thrive and the poor ones decline. He outlined how this geography is tied to ecological issues and systems of financing:

> *Proper ecological balance will not be found without directly addressing the structural forms of poverty* that exist throughout the world.
>
> Rural poverty and unjust land distribution in many countries, for example, have led to subsistence farming and to the exhaustion of the soil. Once their land yields no more, many farmers move on to clear new land, thus *accelerating uncontrolled deforestation*, or they settle in urban centers which lack the infrastructure to receive them.
>
> Likewise, some *heavily indebted countries are destroying their natural heritage, at the price of irreparable ecological imbalances*, in order to develop new products for export....

...it would be wrong to assign responsibility to the poor alone for the negative environmental consequences of their actions. Rather, the poor, to whom the earth is entrusted no less than to others, must be enabled to find a way out of their poverty. This will require a courageous reform of structures, as well as new ways of relating among peoples and States

WORLD DAY OF PEACE 1990 MESSAGE "PEACE WITH GOD THE
CREATOR, PEACE WITH ALL CREATION," JANUARY 1, 1990

He further advised that ecologically and technologically advanced countries must offer advisors, education, and technical training to less developed nations so they can "learn how to fish" (or invent their own new methods that fit their particular fish and waters). Yet if the developed nations and World Bank *do not also forgive the crushing debts of vulnerable developing countries*, these nations will feel forced to neglect their people while denuding their natural resources to produce goods for multinational corporations to pay their debts and interest, hastening the cycle of increasing poverty, debts, and dependence. The pope told President George W. Bush in 2001:

Respect for nature by everyone, a policy of openness to immigrants, the cancellation or significant reduction of the debt of poorer nations, the promotion of peace through dialogue and negotiation, the primacy of the rule of law: These are the priorities the leaders of the developed nations cannot disregard.

ADDRESS TO PRESIDENT GEORGE W. BUSH AT CASTEL GANDOLFO, JULY 23, 2001

Financial Systems Must Be Morally Reformed for Work, Dignity, and Shared Gain

The popes who have followed John Paul II have certainly affirmed his recommendations. In 1993, John Paul had established the Centesimus Annus— Pro Pontefice Foundation to study how to practically apply Christian social justice principles in development, experimenting with the pope's own projects funded through gifts and donations. Honoring the twentieth anniversary of its founding, with an international conference entitled "Rethinking

Solidarity for Work: Challenges of the 21st Century," Pope Francis sent shock waves through conference visitors and the press when he stated that nothing less than a wholescale value restructuring of economic markets and financial operations would be sufficient to achieve real solidarity. He emphasized solidarity must no longer be conceived

> ...as simple assistance to the poor but as a global rethinking of the entire system, seeking ways to reform and correct it in a manner consistent with fundamental human rights, the rights of all men and women. This word 'solidarity,' which isn't seen in a good light by the economic world...needs to have its deserved social citizenship restored.
>
> Chasing the idols of power, profit, and money over and above the value of the human person has become a basic rule of operation... It has been forgotten, and still we forget, that above business logic and the parameters of the market lies the human being and that there is something owed to humans as humans, in virtue of their profound dignity: the opportunity to live in dignity and to actively participate in the common good.
>
> ADDRESS TO CENTESIMUS ANNUS—PRO PONTEFICE FOUNDATION, IN ROME, MAY 25, 2013

Pondering global economics is not new for Pope Francis. When he was a cardinal, he cowrote a book in dialogue with his friend Rabbi Abraham Skorka called *On Heaven and Earth* (2010). In it, they called for forging "a moral vocabulary on economics," and the cardinal observed that "the globalization that makes everything uniform is essentially imperialist and instrumentally liberal, but it is not human. In the end, it is a way to enslave the nations..."

He cited an example of the evils of the present globalization of markets:

> Someone who operates a business in a country and then takes that money to keep it outside of the country *is sinning* because he is not honoring with that money the country to which he is owes his wealth, or the people that worked to generate it.

Thus the outsourcing of jobs and the taking of profits from businesses in one country and putting them in tax-sheltered accounts in another is nothing short of cold evil and sin. The money must be kept in the community, he said, moving through it for the good of all, not skimmed off and out, or locked down in austerity programs that stifle the economy from building and providing opportunities for work and living.

As pope, Francis has favored neither capitalism nor communism but God-oriented humanism that offers "work, work, work," as he told Sardinian workers cut from their jobs because of austerity programs. The international community, governments, and businesses must serve the people and God, seeking opportunities for all people to develop their potential; imposing reasonable and responsible limits on all rights in order to foster the common good; establishing fair, living wages with shared profit and decision making in companies; and shared use and protection of common natural resources.

At Mass at the Cagliari Cathedral, he preached boldly that the recession and austerity programs were not Italy's or Europe's fault per se:

> It is the consequence of a world choice, of an economic system that brings about this tragedy, an economic system that has at its centre an idol which is called money…
>
> We don't want this globalised economic system which does us so much harm. Men and women have to be at the centre (of an economic system) as God wants, not money."

HOMILY, MASS IN CAGLIARI, SARDINIA, SEPTEMBER 22, 2013

New Economic Measurements of Health, Progress, and Success

Popes John Paul II, Benedict XVI, and Francis have all called for innovations in financial structures and how the systems, nations, and businesses define and calculate success and progress. In an address to the UN Food and Agriculture Organization in 1989, the pope called for a reassessment of the concept of "growth" beyond just profits to include people and ecology:

The relationship between problems of development and ecology also demands that economic activity project and accept the expenses entailed by environmental protection measures demanded by the community—be it local or global—in which that activity takes place.

Such expenses must not be accounted as an incidental surcharge, but rather as an essential element of the actual cost of economic activity. The result will be a more limited profit than was possible in the past, as well as the acknowledgment of new costs deriving from environmental protection. Those costs must be taken into account both in the management of individual businesses and in nation-wide programs of economic and financial policy, which must now be approached in the perspective of regional and world economy.

In the end, we are called to operate beyond narrow national self-interest and...defense of the prosperity of particular groups and individuals. These new criteria and costs must find their place in the projected budgets of programs of economic and financial policy for all countries, both the developed and the developing.

ADDRESS TO THE XXV SESSION OF THE CONFERENCE OF
THE UN FOOD AND AGRICULTURAL ORGANIZATION, NOVEMBER 16, 1989

New forms of economics guided by morals and environmental awareness—such as compassionate capitalism, natural capitalism, and sustainable capitalism—are all growing movements. A tool for full-cost public-sector accounting called the Triple Bottom Line (TBL or 3BL) method is spreading internationally and being recognized and promoted by the UN. It tabulates the three core elements of Profit, People, Planet. Visualized as three intersecting circles, TBL measures economic, social, and ecological success. Even the U. S. Environmental Protection Agency is putting forth new planning and measuring tools for building a "new economy." It is proposing a visionary path that seeks to create "Equitable, Healthy, and Sustainable Communities" using the TBL.

Economists have developed other national indicators like the Adjusted GDP (that amortizes natural resource losses), the Happy Planet Index, the Index of Sustainable Economic Welfare, and the EcoBudget (which measures the ecological footprint). Now nations have to start adopting these tools as standard practice and letting them influence their planning and policies. We think and therefore we are; we imagine, plan and measure our progress and thus we become.

The popes' teachings push nations, corporations, businesses, and stockholders to ask themselves: Can we reduce expectations on earnings and growth to be more sustainable, healthy, and, frankly, less greedy? Instead of sending out quarterly earnings, which put all the attention on short-term profits, should corporations and shareholders return to annual reports and more long-range planning? These are questions that corporate boards, leaders, economists, and accountants are grappling with to help us reorient ourselves and our economic actions toward having a positive effect on the planet and ensuring a hospitable, beautiful one in the future, where all people will have access to dignified work and ways to make and enjoy a living.

The Role of Responsible, Faith-Filled Shareholders

Shareholders are starting to understand their role in creating this better world too. One financial investment company uses the advertising line: "Your life has your fingerprints all over it. Why should the way you invest be any different?"

How true. When deciding where to invest, stockholders need to consider the full range of a company's products, processes, worker-to-corporate officer pay percentages, and environmental and social justice criteria. Choosing an investment strategy based on doing good as well as making money is an embodiment of Jesus's words, "Where your treasure is, your heart is also"—or, expressed more crassly, "Put your money where your mouth is."

This is the reasoning behind the over 840 socially responsible investment funds around the globe, which are performing similarly to regular funds, with some doing better. Among these funds are Parnassus, Pax World, iShares Select Index, Calvert Equity Income Fund, Walden Equity Fund, Appleseed Fund, TIAA-CREF Social Choice Equity Fund, and Domini Social Equity Fund.

Amy Domini, founder of Domini Social Investments, received an honorary degree from Yale Divinity School in 2007. In her acceptance speech, she said:

For me, finance and capitalism are not exempt from God's will. Investors are not innocent in the actions of the teams they enable. Right is right and wrong is wrong.

I am as forgiving a Christian as any but there are ways to make money that go beyond the reach of decency. Further, I found that when I let my "Jesus time" out of the Sunday morning box and into my career, I became a powerful force for good.

Not Just Short-Term Charity, But Long-Term Solidarity and Fraternal Partnerships

When a person or community is in a short-term crisis, charity is needed. But when long-term changes are required for stability, improvement, growth, or healing, then training or education, partnerships, and resources are needed. Nobel Peace Prize-winning economist and banker Muhammad Yunus has been promoting microloans, microfinancing, and community banking in Bangladesh as ways to reform some of the values of the banking systems and assist people to grow out of debt, poverty, and dependence. Like the popes, he advances a concept of financing that cares about a full range of values and freedoms that help people develop their potential, so they can participate in and give back to the community at large. His loans have a far greater repayment rate than traditional banks. He promotes "inclusive growth," including increasing income to the poor as a necessary criteria for real development. Banking cannot be just about profits for the stockholders and the wealthy few.

In *The End of Poverty*, international development specialist Jeffery Sachs touts many successful examples of development and environmental restoration going hand in hand. Villagers lay out their needs and suggested solutions and work with experts to help accomplish them. The local people often say to the outsiders, your ideas are fine, but first we need a well, or first we need to plant trees, or first we need to fix this road or bridge so we can get the materials. And they are right.

In such community-based conservation, local people are not only participants, but decision-makers and laborers. These projects are designed to:

- involve the local people and empower them;
- raise the personal standard of living and quality of life for individuals and the community;
- raise the education level as people learn new skills and attitudes;
- increase biodiversity, habitats, moisture, fertility, and beauty (by building integrated natural communities);
- strengthen human communities, uniting people in teamwork and community pride as their environment improves;
- build democracy and citizen involvement—engaging people politically, for as they care for their physical surroundings, they begin to have a voice.

Pope John Paul II so believed in these principles that he designed his Sahel Foundation to promote them. All funded projects must be proposed by the people themselves, with the foundation supplying the training, technology, and materials. Five years after the foundation was launched, John Paul told the UN Environment Program,

> Even though the project seems small and inadequate in the face of such vast needs, nonetheless it is a concrete effort to help the people there and to contribute in some degree to the future of the African continent, a future which ultimately rests in the hands of the African peoples themselves.

> ADDRESS TO THE MEMBERS OF THE AGENCY OF THE
> UNITED NATIONS, NAIROBI, KENYA, AUGUST 18 1985

Models of Ecological Development and Hope

The Green Belt Movement, of course, is an example of this kind of partnership, but there are many other successful examples spread out far and wide through communities across the globe. For instance, the mission of Heifer International is "to work with communities to end hunger and poverty and care for the Earth." Through this nonprofit organization, people in need get gifts of livestock, chicks, bees, and trees; families raise them and then pass on a portion of the offspring to someone else as a gift to help them start on a new life too. Its mission embraces the 12 Cornerstones:

- fostering full participation
- addressing genuine need and justice
- offering education and training
- establishing accountability
- promoting sharing and caring
- enhancing nutrition and income
- encouraging sustainability and self-reliance
- improving animal management
- supporting vulnerable women and families
- improving the environment
- deepening spirituality
- passing on the gift

How much they sound like John Paul II's guiding principles!

Recipients are trained in how to raise the animals and plants sustainably, nurturing the earth while nurturing their families, their communities, and their ecosystems for the benefit of all. Heifer was founded by Dan West, a farmer in the American Midwest who was a member of the Church of the Brethren.

Habitat for Humanity is a nonprofit, ecumenical Christian housing organization with a similar approach, requiring those who receive housing to work on building their own home and then help others. The houses are constructed in affordable and sustainable ways that fit the locale, as in the new Habitat EcoVillage in River Falls, Wisconsin. The project legitimately boasts: "Now with nearly a year under our belts and the first families ready to move in to their net-zero homes, we are proving that sustainable housing can and should be beautiful and affordable for all." (Even in places with below-zero winters!)

Building on similar partnership principles, the Earth Institute at Columbia University, directed by Jeffrey Sachs, has been heading up the Millennium Villages Project (MVP). The MVP works with communities of 5,000 to 70,000 citizens in five key interconnected areas: agriculture, health, education, business development, and infrastructure (including water and sanitation). All community projects and innovations are locally driven, sustainable and earth friendly, engaging men and women equally.

Dr. Paul Farmer is taking a similar community and personal engagement approach in medicine with his Partners in Health program, training community health workers and sending them out for home health visits and education. He started the program in response to the extreme poverty and lack of health care in Haiti, and it has been so successful he has been helping spread it to other countries.

All these nonprofit organizations and many more around the world have stories of successful partnerships that multiply success by passing it on, creating pathways of hope. Just as each person is endowed by the Creator at birth with an ecological vocation, each is also endowed with the right to fair access to the garden of this planet and its fruits. And each has a vocation of service, to act as a channel of God's love and abundance, passing on gifts to others. For in service to others, there is the joy of meaning and an even deeper sense of worth, actually discerning one's useful place in the universe. These organizations have given us models of how to structure workable development, changing our paradigm to fit that of the Creator's original design. John Paul II said:

> Every person, every people, has the right to live off the fruits of the earth. At the beginning of the new millennium, it is an intolerable scandal that so many people are still reduced to hunger and live in conditions unworthy of humans.
>
> We can no longer limit ourselves to academic reflections: we must rid humanity of this disgrace through appropriate political and economic decisions with a global scope.
>
> JUBILEE OF THE AGRICULTURAL WORLD ADDRESS, VATICAN CITY, NOVEMBER 11, 2000

The condition of women must also be improved
as an integral part of the modernization of society.

ADDRESS TO THE PONTIFICAL ACADEMY OF SCIENCES, STUDY WEEK ON
RESOURCES AND POPULATION, NOVEMBER 22, 1991

6. Oppression and Exclusion of Women and Others

In spring 2013, the World Health Organization released the first international report on violence against women, proclaiming it a "global health epidemic." Averaged out, one third of all women in the world are victims of sexual or physical violence—mostly inflicted by husbands, boyfriends, fathers, brothers, and other men in their lives. According to this report, 38% of murdered women are killed by their intimate partners. This domestic abuse is doubly damaging in that it is perpetrated by the people who are supposed to love them. Sharing a living space with their abusers means women, especially mothers, can be imprisoned in their own homes and have their children kept from them or threatened with harm as well.

Verbal and physical abuse is a tragic, everyday domestic occurrence, but then there is rape, which is not only a common crime, but is practiced often as a weapon of war. Rape is rampant in refugee or resettlement camps, as well as in "peaceful" cities and towns. Finally, there is the sexual slavery trade. Along with all these types of sexual violence come the added burdens of sexually transmitted diseases, unsought pregnancies, emotional issues, and often alcohol and drug abuse to cope with the violence.

Half the Sky

Journalists Nicholas Kristof and Sheryl WuDunn documented some of the stories and statistics from around the world in their best-selling book *Half the Sky: Turning Oppression into Opportunity for Women Worldwide*. The book describes the global epidemic of female oppression through stories of real girls around the world, and the heroic women working to help them achieve better lives through freedom, education, and empowerment. They address not only the kidnappings and harsh, unchosen lives of the girls in brothels and sex trafficking, but also the genital mutilation and death in childbirth many suffer.

In an accompanying documentary (shown on PBS and international screenings), American actresses Meg Ryan, Eva Mendes, and others visit some of the girls and talk with their liberators. Together with the book, the film *Half the Sky* has sparked a movement. Author Stuart Perrin, who has been involved in rescue work that has freed some girls from brothels, brought these situations to dramatic life in his novel *Little Sisters*. Works like these, together with horrific news stories like that of the 2012 gang rape and death of a young woman on a New Delhi bus have finally outraged the world community enough to draw global attention to these problems. But the attention and decisive action are long overdue.

Back in 1995, in anticipation of the UN Conference on Women in Beijing, Pope John Paul II focused his World Day of Peace Message on "Women as Makers of Peace" to honor women and the need for equality in treatment and opportunities. He observed:

> ...many women, especially as a result of social and cultural condition-
> ing, do not become fully aware of their dignity. Others are victims
> of a materialistic and hedonistic outlook which views them as mere
> objects of pleasure, and does not hesitate to organize the exploitation
> of women, even of young girls, into a despicable trade. Special concern
> needs to be shown for these women, particularly by other women
> who, thanks to their own upbringing and sensitivity, are able to help
> them discover their own inner worth and resources.

Women like these are the courageous rescuers in *Half the Sky*. John Paul wrote a more comprehensive global message on the topic entitled "Letter to Women," where among other things, he urged the world to take violence against women seriously and act to prevent it:

When we look at one of the most sensitive aspects of the situation of women in the world, how can we not mention the long and degrading history, albeit often an "underground" history, of violence against women in the area of sexuality? At the threshold of the Third Millennium we cannot remain indifferent and resigned before this phenomenon. The time has come to condemn vigorously the types of sexual violence which frequently have women for their object and to pass laws which effectively defend them from such violence.

Nor can we fail, in the name of the respect due to the human person, to condemn the widespread hedonistic and commercial culture that encourages the systematic exploitation of sexuality and corrupts even very young girls into letting their bodies be used for profit.

He mentioned the violent injustice of rape, in war zones as well as in places of prosperity and peace. In this letter, he did not mourn women as mere victims but applauded their acts of caring and triumph over adversity to get the rights they deserve:

Here I cannot fail to express my admiration for those women of good will who have devoted their lives to defending the dignity of womanhood by fighting for their basic social, economic and political rights, demonstrating courageous initiative at a time when this was considered extremely inappropriate, the sign of a lack of femininity, a manifestation of exhibitionism, and even a sin!

He encouraged all humanity to support the journey of educating women, saying how progress for "women's liberation" has been difficult, but the benefits to all humanity are and will be worth all the trials and mistakes: "The journey must go on!"

In 2002, John Paul again spoke out vehemently (as he had many other times before and after) for immediate policies and actions to crack down on the worldwide criminal business of kidnapping, selling, and trafficking of human beings. He asked governments and international entities to create "juridical instruments to halt this iniquitous trade, to punish those who profit from it, and to assist the reintegration of its victims," and stop the overly sexualized media views of women:

> The disturbing tendency to treat prostitution as a business or industry not only contributes to the trade in human beings, but is itself evidence of a growing tendency to detach freedom from the moral law and to reduce the rich mystery of human sexuality to a mere commodity.
>
> Attention needs to be paid to the deeper causes of the increased "demand" which fuels the market for human slavery and tolerates the human cost which results. A sound approach…will lead also to an examination of the lifestyles and models of behaviour, particularly with regard to the image of women, which generate what has become a veritable industry of sexual exploitation in the developed countries. Similarly, in the less developed countries from which most of the victims come, there is a need to develop more effective mechanisms for the prevention of trafficking in persons and the reintegration of its victims.
>
> LETTER TO ARCHBISHOP JEAN-LOUIS TAURAN ON THE OCCASION OF THE INTERNATIONAL CONFERENCE ON "TWENTY-FIRST CENTURY SLAVERY: THE HUMAN RIGHTS DIMENSION TO TRAFFICKING IN HUMAN BEINGS," MAY 15, 2002

Society and the media must take responsibility for their encouragement of objectification of women, and even children. The act of turning girls, boys, and women into objects to be sexually used and thrown away is the consumer culture of waste personified. The victims must be cared for and not blamed for the actions against them.

Pope Francis, too, has long been a staunch soldier in the fight against sexual trafficking, visiting victims and working on Argentinian initiatives against the trade. On September 23, 2011, as cardinal, he said a Mass for the victims at the Buenos Aires Plaza Constitucion in honor of the International Day Against Sexual Exploitation and Human Trafficking. The Mass marked nearly one hundred years since the first law in the world to outlaw child prostitution, enacted in Argentina. Francis reprimanded the city:

> For many, Buenos Aires is a meat grinder which destroys their lives, breaks their will, and deprives them of freedom...
>
> ...there are people committing human sacrifice, killing the dignity of these men and these women, these girls and boys who are submitted to this treatment, to slavery. We cannot remain calm.

In the first summer of his papacy, he issued a *motu proprio,* a legislative initiative updating Vatican laws to assign criminal prison penalties for sexual acts with and violence against children, child prostitution and trafficking, and child pornography, strengthening the larger law against child violence. This act is part of his work of reform not just for the Vatican City state, but the Church as a whole, and the world. Pope Francis has shown that he is well aware that there is more to do to stop sexual abuse within the Church, and hold the hierarchy of the Church and clergy accountable.

The problems of relationships between clergy and women or adolescent girls is a still-unscrutinized part of the picture. These kinds of settlements happen more quietly, beyond the eyes of the press. Women don't want to be placed in the stereotype of the "temptress," grilled, or blamed for the wrongs done to them.

The Girl Effect

The rampant international violence against women and children alters human ecology in its most profound sense. Consider how this kind of sexual exploitation and violence—suffering it, witnessing it and enduring its repercussions—affects families and children. Such a culture of waste teaches new generations of boys and young men that women don't deserve to be treated with dignity and respect. And it teaches girls to be apprehensive about participating in society, diminishing their sense of worth, dignity, competence, and dreams.

Thus, we are wasting the creativity, intelligence, and loving energies of this large portion of the humanity God created—"half the sky," as Kristof and WuDunn dubbed the offerings of women. Without the full energies of humanity, we will not be able to meet the broader ecological crises we face.

In contrast, what can education and women's participation and leadership in society do?

It can transform it. As Pope John Paul II said and development agencies have proven, the ideas and energies of women can help turn a society around, particularly in the area of health, education, the environment, and the economy. A very brief but powerful YouTube video called "The Girl Effect" shows, with simple animation and a statistic or two, how effective educating women can be in transforming a community. Getting an education tends to delay childbirth to an age when women are more able to make discerning choices about their futures and those of their children, which has positive repercussions on population, childcare, health, and resources. Women with an education are more likely to invest in education for all their children, be more knowledgeable about health issues and active civic decision-making, and be more entrepreneurial.

Every development study shows that educating women is the best dollar-for-dollar expenditure to uplift a community. Educated women can offer powerful leadership, often with different values and perspectives than men. They tend to serve others and aim for the common good instead of just profits. John Paul II emphasized this in his "Letter to Women":

> Doubtless one of the great social changes of our time is the increasing role played by women...in an executive capacity, in labor, and the economy. This process is gradually changing the face of society, and it is legitimate to hope that it will gradually succeed in changing that of the economy itself, giving it a new human inspiration and removing it from the recurring temptation of dull efficiency marked only by the laws of profit.
>
> "EQUAL OPPORTUNITY STILL URGENTLY NEEDED," MESSAGE DELIVERED ON AUGUST 20, 1995

The Pro-Women Pope (To a Point)

Because of the pope's stance against the ordination of women as Roman Catholic priests, this part of his legacy his little-known. The pope enthusiastically promoted equal education, access, pay, and leadership in society for girls and women—not only for justice but as *significant factors to improve the ecological, economic, and human health of the world.* Addressing the Beijing UN Conference on Women, he wrote:

As far as personal rights are concerned, there is an urgent need to achieve *real equality* in every area: equal pay for equal work, protection for working mothers, fairness in career advancements, equality of spouses with regard to family rights and the recognition of everything that is part of the rights and duties of citizens in a democratic State.

This is a matter of justice but also of necessity. Women will increasingly play a part in the solution of the serious problems of the future: leisure time, the quality of life, migration, social services, euthanasia, drugs, health care, the ecology, etc.

In all these areas, a greater presence of women in society will prove most valuable, for it will help to manifest the contradictions present when society is organized solely according to the criteria of efficiency and productivity, and it will force systems to be redesigned in a way which favors the processes of humanization which mark the "civilization of love."

"LETTER TO WOMEN," JUNE 29, 1995

Six years after Beijing, Pope John Paul II reiterated his commitment to women's education and leadership by sending a representative of the Holy See, Joan McGrath-Triulzi, to make these comments to the follow-up conference on gender equality:

The Holy See would like to reiterate its support for what it called "the living heart" of the Beijing Platform for Action: the recognition of the dignity of women, the importance of strategies for development, ending violence against women, access to employment, land and capital, and the provision of basic social services. . . .

Everyone can see that Catholic schools, hospitals, and humanitarian agencies all over the world have responded seriously to the pope's exhortation. The Holy See continues to be a major provider of basic social services to girls and women, especially in developing countries. . . .

Lastly, investment in basic social services is the bedrock for women's wellbeing and economic development. To be actors in the changing economy, women need to be physically and mentally healthy and possess marketable skills. It is imperative then, that the education and health of girls and women be a priority in development programs.

MESSAGE FROM THE HOLY SEE AT THE UNITED NATIONS GENERAL ASSEMBLY, "WOMEN 2000: GENDER EQUALITY, DEVELOPMENT, AND PEACE FOR THE 21ST CENTURY," OCTOBER 19, 2001

John Paul II had already put these social principles into practice by making the education of girls a priority in his Foundation for the Sahel and in the foundation he created for development in the barrios of Latin America. He advised that though all religions and cultures have to be respected, the global community has an even more integral obligation to provide safety, individual freedom, and human rights for women across all cultures and religions.

Malala's Courageous Example

Malala Yousafzai, the Pakistani teenager who was shot in the head by the Taliban for going to school, is a hero of this truth. For her first public appearance since her recovery, she chose to stand before the United Nations on her sixteenth birthday and plead for freedom from terrorism in the name of God and free, compulsory access to quality education for *every one of the world's* children:

> I speak—not for myself, but for all girls and boys. I raise up my voice—not so that I can shout, but so that those without a voice can be heard. Those who have fought for their rights: Their right to live in peace. Their right to be treated with dignity. Their right to equality of opportunity. Their right to be educated . . . I am focusing on women's rights and girls' education because they are suffering the most.

Courageously, she said that the shooting transformed her:

> The terrorists thought that they would change our aims and stop our ambitions but nothing changed in my life except this: Weakness, fear, and hopelessness died. Strength, power, and courage was born.

The YouTube recordings of her dynamic UN speech are so compelling for two reasons: because viewers can encounter this smiling, courageous teenager, and because they can see that one young woman—and all young women (and men) unleashed from the bonds of violence and oppression—can make brave stands to change the future. Deeply rooted in her faith, this teen had the guts to say to the Taliban in front of every nation on earth not only that they were wrong in their violent actions *but in their theology.* This confidence in the face of real danger almost outstrips belief. Their goals are not of God, she said, and the Taliban should be ashamed in front of the Almighty, the religion of Islam, and the world:

> They think that God is a tiny, little conservative being who would send girls to the hell just because of going to school. The terrorists are misusing the name of Islam and Pashtun society for their own personal benefits. Pakistan is peace-loving democratic country. Pashtuns want education for their daughters and sons. And Islam is a religion of peace, humanity, and brotherhood. Islam says that it is not only each child's right to get education, rather it is their duty and responsibility.

A devout Muslim, she refused to consider that revenge or violence would help heal her, or the world—only education and learning for all people and all religions would pave the way to a better future:

> I do not even hate the Talib who shot me. Even if there is a gun in my hand and he stands in front of me, I would not shoot him. This is the compassion that I have learned from Muhammad, the prophet of mercy, Jesus Christ, and Lord Buddha. This is the legacy of change that I have inherited from Martin Luther King, Nelson Mandela, and Muhammad Ali Jinnah. This is the philosophy of nonviolence that I have learned from Gandhi Jee, Bacha Khan, and Mother Teresa. And this is the forgiveness that I have learned from my mother and father. This is what my soul is telling me, be peaceful and love everyone.

She asked all people and all nations to rise up together against such poor reflections of God—making God narrow, mean and exclusive—not only in terrorist Islamic groups but in all religions, countries, and communities on earth. Malala said:

> So today, we call upon the world leaders to change their strategic policies in favour of peace and prosperity. We call upon the world leaders that all the peace deals must protect women and children's rights. A deal that goes against the dignity of women and their rights is unacceptable. We call upon all governments to ensure free compulsory education for every child all over the world. We call upon all governments to fight against terrorism and violence, to protect children from brutality and harm.
>
> We call upon the developed nations to support the expansion of educational opportunities for girls in the developing world. We call upon all communities to be tolerant—to reject prejudice based on caste, creed, sect, religion, or gender. To ensure freedom and equality for women so that they can flourish. We cannot all succeed when half of us are held back. We call upon our sisters around the world to be brave—to embrace the strength within themselves and realise their full potential...
>
> And if we want to achieve our goal, then let us empower ourselves with the weapon of knowledge and let us shield ourselves with unity and togetherness....Let us wage a global struggle against illiteracy, poverty, and terrorism, and let us pick up our books and pens. They are our most powerful weapons. One child, one teacher, one pen, and one book can change the world. Education is the only solution. Education first.

Women Are the Most Underutilized Ecological Energy Source in the World

As Pope John Paul II taught and Malala proved, to include girls and women in the full embrace of humanity is not just a practical call but one of faith—a call from the God who created all and said they are good, who took the human being in the Bible, the unsexed *adam* of the earth, and split it in two to make man and woman. God made the evolutionary structures of the species and society work best through the equal complementarity of male and female, not through the oppression of women as the property of men.

Pope John Paul II also wanted to make clear that though women and men are absolutely equal in integrity, necessity, dignity, potential, and importance, this did not mean, obviously, that they are biologically, physically, or emotionally the same. He stated that women *in general* have a "feminine genius," an inner natural strength and inclination geared toward relationship-building that can contribute to a "civilization of love":

> Perhaps more than men, women *acknowledge the person,* because they see persons with their hearts. They see them independently of various ideological or political systems. They see others in their greatness and limitations; they try to go out to them and *help them.*"
>
> "LETTER TO WOMEN," JUNE 29, 1995

What he called "the female genius" was not another way to keep women in a limited, defined scope of endeavor. It was, in effect, the opposite, meaning that through societal oppression we have been stifling a valuable source of energy and inspiration that is desperately needed at every level of society. Psychologists have defined the complimentary forms of "genius" between men and women as feminine and masculine qualities that both men and women possess but in different proportions.

Other cultures have noted the differences and described them in varied ways. The writer Alice Walker, in an essay "All Praises to the Pause," mused on the Swa people of the Amazon who have associated certain tendencies with males and females in their societies:

> They tell us that in their society men and women are considered equal but very different. Man, they say, has a destructive nature: it is his job therefore to cut down trees when firewood or canoes are needed. His job also to hunt down and kill animals when there is need for more protein. His job is to make war, when that becomes a necessity.
>
> The woman's nature is thought to be nurturing and conserving. Therefore her role is to care for the home and garden, the domesticated animals and the children. She inspires the men. But perhaps her

most important duty is to tell the men when to stop. It is the woman who says: Stop. We have enough firewood and canoes, don't cut down any more trees. Stop. We have enough meat; don't kill any more animals. Stop. This war is stupid and using up too many of our resources. Stop.

.... when the Swa are brought to this culture they observe that it is almost completely masculine. That the men have cut down so many trees and built so many excessively tall buildings that the forest itself is dying; they have built roads without end and killed animals without number. When, ask the Swa, are the women going to say Stop? Indeed. When are the women, and the Feminine within women and men, going to say Stop?

Time to Say "Enough"

That is one of the key questions of our time. To stabilize the climate and save the ocean, to save the forests and the fresh water, to save endangered species, to stop the waste, to attain a shift in global economics and financial markets toward fraternal sharing and solidarity, some of us are going to have to stand up and say STOP. Enough already. We have *Enough*. Let's stop, and change things so they work better for all. Pope John Paul II firmly believed that Catholics and other Christians can and should lead the way, with women at the front of the pack.

The other voices we all need to hear are those homey practical ones that say, "Let's invite those people over to share our dinner, and let's rearrange our menu so everyone gets something to eat. Let's take care of what we have so we won't have to replace it. Let's share the chore list so everyone gets something to do and some allowance. Let's be educated, all of us, and try new things. Let's make sure we take care of the kids and the grandkids, and let's not forget the grandparents and aunts and uncles. Let's remember to clean up after ourselves. And let's have fun together as a family."

Pope John Paul II restated his call for the need for more women in all levels of leadership and professions to deal with all of today's pressing problems:

How can we fail to see that, in order to deal satisfactorily with the many problems emerging today, special recourse to the feminine genius is essential? Among other things, I am thinking of the problems of education, leisure time, the quality of life, migration, social services, the elderly, drugs, health care, and ecology. "In all these areas, a greater presence of women in society will prove most valuable," and "it will force systems to be redesigned in a way that favors the processes of humanization which mark the "civilization of love."

"EQUAL OPPORTUNITY STILL URGENTLY NEEDED," AUGUST 20, 1995

The Feminine Genius at Work on the Environment

Lest the Pope John Paul II's "feminine genius" be viewed as cliché, or another form of stereotyping, or a sentimental vagary untied to ecological issues, studies do show how women and men often react differently to environmental issues. In a 2003 study by the Institute for Women's Policy Research, researchers Dr. Amy Caiazza and Allison Barrett found:

- Most men and women support increased government spending for the environment. Women, though, are less likely than men to support cuts to environmental spending.
- When it comes to environmental regulation, women are less sympathetic to business than men.
- Women have more positive feelings toward environmental activists than men.
- Both women and men reject "jobs versus the environment" as a false choice.
- Women are particularly concerned about environmental problems that create risks for their health and safety, especially at the local level.
- Women are more interested in environmentalism as a way to protect not only themselves and their families but also others. [The researchers attributed this to "women's higher levels of empathy, altruism, and personal responsibility."]
- Women have less trust that the institutions responsible for protecting the environment are actually doing their jobs.

- Overall, women are less likely to participate politically than men. But women are more likely to volunteer for and give money to environmental causes.
- Women's environmental leadership has been particularly prominent in a few key areas, especially in local environmental movements and in green consumerism.

IWPR PUBLICATION 1913, "ENGAGING WOMEN IN ENVIRONMENTAL ACTIVISM," SEPTEMBER 2003, QUOTED IN SLIGHTLY CONDENSED FORM FROM PP. 1-2.

So there really *is* something measurable to women's connection to the environment. Consider the immense success of women-led grassroots forestry initiatives, such as the Green Belt and Chipko Andolan Movements. The Chipko Andolan ("Embrace the Trees") Movement began in the Himalayas in India in the 1970s, when women wrapped themselves around tree trunks to protect them from contractors. Inspired by this, a similar conservation movement began in southern India, called the Appiko Adolan Movement. This movement also included men and children, and expanded by founding clubs and tree nurseries, and by using folk dances, slide shows, street drama, forest marches, among other things, to the educate citizens about the benefits of conservation and the saving and replanting of forests.

In the Republic of Korea, women finance their Mothers' Clubs by raising and selling trees to supply fuel and reforest the countryside. Women in the Upper Solo Valley of Indonesia, in Jamaica, Thailand, China, and Sudan are working on their own versions of conservation and tree-planting enterprises, restoring the lands around them and improving their family and community economies.

All over North America, women are starting environmental action groups: the Women's Global Green Action Network, the Women's Environmental Institute, Mothers Against Dioxin, Mothers Clean Air Force, the National Stroller Brigade, the White Earth Land Recovery Project, the Indigenous Women's Network, and so many more. At the community level we have ecological and economic restoration projects like Majora Carter's Greening the Ghetto/Sustainable South Bronx, Lily Yeh's Village of Arts and Humanities in North Philadelphia, and Albina Ruiz's *Cuidad Saludable* in Peru, to name just a few.

Is it mere coincidence that Costa Rica, New Zealand, Iceland, and Norway, the globe's four real contenders in the race to be the first truly sustainable countries on the planet, are all headed by women? It has to be noted, of course, that the Vatican and Maldives, radically small-sized contenders, are led by men. Also notable is that Germany, one of the countries furthest ahead in integrating solar into its economy and national grid, is also led by a woman. Chancellor Angela Merkel is a chemist who was formerly the minister of Women and Youth, and then the minister of Environment, Nature Conservation, and Nuclear Safety. Bhutan has asked Dr. Vandana Shiva to help it become the first "100% organic nation."

Not bad for the feminine genius and "women's work." Imagine the restorative energy that would be unleashed if the world could fully put into play all the creative energies and leadership of the underutilized populations of women. The key to opening this door has been education, health, respect, and access to resources, authority, and decision-making. In his "Letter to Women," John Paul II offered his heartfelt gratitude: "Thank you, *women who work!* You are present and active in every area of life—social, economic, cultural, artistic and political."

The pope admitted that "we are heirs to a history which has *conditioned* us to a remarkable extent" against the inclusion of women, stealing away women's voices, their power over their own opinions and views, and their ability to contribute to humanity as a whole. This has to be overturned, he said, for the good of all.

But even as women choose to work and lead lives outside the home, he admonished that their loving and caring role within their home should never be diminished, but should be admired and supported, with the burdens shared by men and others in the family and society.

Multiplying the Ecological Benefits: Women's Groups and Loans

All studies show that the pope was right. The world's ecological and development problems *cannot* be solved without the feminine genius. Women now make up 70% of the world's poor. Environmental problems hit them first and affect them most harshly, followed by their children, especially their daughters. In the less industrialized areas of the globe, women are the ones who gather firewood and haul water, plant and gather the food, cook, care for the health and welfare of the children, and get them ready for school, and then care for their husbands, aging parents, and other family members. They prepare the dead for burial, cook and serve at funerals, and oversee births and weddings.

When firewood or water is far away, they spend additional hours walking and carrying, and in some areas, often under threat of rape, kidnapping, or death. Women and girls have higher rates of illness and malnutrition. Many do not have an education or a voice in their own marriages, families, or communities. Yet all this can be turned around with just a little investment in them.

In 2000, in a small Nepalese village, thirteen women received two sheep and a ram, or two goats and a buck, along with training in sustainable development and animal management from Heifer International. They were also offered simple classes in reading, writing, and running group meetings and activities.

They called themselves the "Parijat Women's Group," or women's awareness group. The livestock added the equivalent of about two hundred dollars annually to each woman's family income. They used this money to improve their families' nutrition and health, buy school uniforms, and send their daughters as well as their sons to school. Some also invested in land or developed small businesses. Beyond this, each member offered a small, assigned monthly contribution to the group's treasury. They used the conglomerate funds to accomplish group projects, such as improving the village road, buying livestock so that other women could join the group, or providing loans to others.

The women gleam with pride and admit that their group "has become famous" in the region. Sarada Bista, the chairwoman of the Parijat Women's Group, said that before the group began, "I could not speak." Another member elaborated, "We have found our voices. We can use our own names. We get mail in our own names; before we could only get it through our husbands. Even though we are poor and did not have a dowry, we can help our husbands."

This story may sound unusual, but when women of little means and education are given a chance, these success stories abound. As Pope John Paul

II stated, investments in women yield a cascade of benefits to a community. In fact, among the most exciting findings of Muhammed Yunus' Grameen Bank is how transforming to communities small loans at low interest to women can be, especially when strengthened by a women's support group or network. Studies show that women are four times more likely than men to spend their income to improve the nutrition, education, and health of their children and family members, thus improving the stability of communities. They have extremely high repayment rates (far higher than men), with defaults extremely rare. As necessity *is* the mother of invention, they tend to be practical and ingenious, and are also the most open to working to improve the environment around them.

The Feminine in Faith

In many cultures of the world, the earth itself is associated with the feminine, as in "Mother Earth." In Judeo-Christian Scripture, the spirit of the Lord that moves over the earth with life is denoted by the Hebrew word רוח (*rua'h*), a noun that is grammatically feminine, meaning wind, breath, inspiration, and spirit. Pope John Paul II highlighted some of the feminine and maternal images of God in the Book of Isaiah 66: 11-14, when the Lord comforts Jerusalem like a mother to a child:

> "they will be carried upon her hip, and dandled upon her knees." And this motherly tenderness will be the tenderness of God himself: "As one whom his mother comforts, so I will comfort you". Thus, the Lord uses a maternal metaphor to describe his love for his creatures.
>
> We can also read an earlier passage in the Book of Isaiah that gives God a maternal profile: "Can a woman forget her sucking child, that she should have no compassion on the son of her womb? Even though these may forget, yet I will not forget you" (Is. 49: 15).
>
> GENERAL AUDIENCE AT CASTEL GANDOLFO, JULY 16, 2003

The emotional, philosophical, and cultural linkages between love, nature, and the feminine are powerfully moving. But they are not always considered by society as important as measurable elements, such as finances or scientific evidence. There are definite connections among societal attitudes: the more a culture disrespects and abuses the earth, the more it tends to do the same to women and children, and vice versa. Violence against women and children often result when they, like livestock and land, are considered legal "property" that can be treated as "needed" or "desired" by men. This is a degrading view of land, animals, women, children, *and* men. That is why movements for gender equity and ecological respect have multiplying positive effects, putting people in right relationships on many levels. It helps men as well as women, allowing men to nurture and feel the joys of children and family and nature.

What About the Catholic Church?

One could then fairly ask: If Pope John Paul II believed so completely in the feminine Spirit within God and the need for the women's education and equal opportunities and leadership, why did he not desire more of this energy for his own Church, so it could lead in this ecological transformation?

In many ways, he did. In his "Letter to Women," he admitted and apologized for the oppression and discrimination against women that the Church had historically engaged in. Then, in his apostolic exhortation to bishops in America in 1999, he stated that "the Church...'denounces discrimination, sexual abuse and male domination as actions contrary to God's plan.'"

The pope sent Mary Ann Glendon, professor of law at Harvard University Law School, to represent the Holy See at the Beijing Conference, and she summed up some of the ways Christianity and Roman Catholic systems have fostered women's progress:

> Doesn't it boggle the mind to realize that the Church succeeded in gaining wide acceptance for the novel ideal that marriage was indissoluble—in societies where men had always been permitted by custom to put aside their wives? That she fostered the rise of strong, self-governing orders of women religious in the Middle Ages? That she pioneered women's education in countries where most other institutions paid scant attention to girls' intellectual development?
>
> Who runs the second largest health care system in the world? Has it not long been managed almost entirely by dynamic Catholic

women executives (mainly religious sisters)? Who runs the world's
largest system of private elementary and secondary education? Has
it not long been largely run by Catholic women, religious and lay, as
teachers, principals, and superintendents?

The pope well understood that for the most part *women had done this work
for themselves*, and that male clergy were not always respectful or supportive precisely because of their isolation and assumption of superiority. There was a time
in European history when most lay people, both male and female, were illiterate
and uneducated. Being illiterate, they were barred from roles of leadership in
the diocesan and Vatican structures of the Catholic Church. An additional bias
was laid against women as being temptresses without brains (especially after St.
Augustine of Hippo wasted his youth chasing after women and *then blamed them*
for it). For centuries, women in Western civilization were generally oppressed
and prevented from getting rights or an education *unless they joined a religious
order in a convent*. Then, within the community of sisterhood, women were given
all the education they were denied in the outside world (just as poor men had
access to education by joining a religious order or the clergy).

Over the centuries, the educated orders of sisters started schools to
educate girls, lifting up women to opportunities they would never have had,
as in the case of Wangari Maathai. So in his 1995 "Letter to Women," John
Paul II honestly apologized for the male domination and added:

> May this regret be transformed, on the part of the whole Church, into
> a renewed commitment of fidelity to the Gospel vision. When it comes
> to setting women free from every kind of exploitation and domina-
> tion, the Gospel contains an ever-relevant message that goes back to the
> *attitude of Jesus Christ himself*. Transcending the established norms of his
> own culture, Jesus treated women with openness, respect, acceptance
> and tenderness. In this way he honored the dignity that women have
> always possessed according to God's plan and in his love. As we look to
> Christ at the end of this Second Millennium, it is natural to ask our-
> selves: how much of his message has been heard and acted upon?"

Of course, this question still hangs in the air.

Looking for Answers

In this, Pope John Paul II was an enigma. As a young university professor, he demonstrated great respect for women, teaching them as equals and inviting them along with the men on the outdoors trips he sponsored. As a cardinal, he included women staff members (very radical!) in the Second Vatican Council sessions he attended. Then, as pope, as parishes were losing priests, he encouraged sisters to be established as parish administrators, building a force of women decision-makers coming up through the parish ranks.

At the top, he added *the first women* to the esteemed ranks of the influential International Theological Commission: Sister Sara Butler of Chicago's University of Saint Mary of the Lake (and a one-time supporter of women priests) and Barbara Hallensleben of Fribourg University in Switzerland. He also appointed the first woman to be president to the Pontifical Academy of Social Sciences (unheard of!)—Mary Ann Glendon.

On the other hand, in contrast to this strong talk and actions, he made it quite clear in his 1994 apostolic letter, *Ordinatio Sacerdotalis* (On Reserving Priestly Ordination to Men Alone), his 1995 "Letter to Women," and his 1988 apostolic letter, *Mulieris Dignitatem* (The Dignity of Women) that for as long as he was pope, ordination for women was out of the question. In this, he was considered a hardliner.

New Possibilities of Ministry and Action

It seemed that Pope John Paul II wanted to open every door to women he could *within the range of established canon law and traditional service structures.* But beyond that line, he would not cross, and during his papacy he wanted to shut the door on a controversy that was splitting the Church. In *Ordinatio Sacerdotalis,* he declared that since Jesus didn't make females part of his group of twelve apostles, then in the pope's ministry of confirming the brethren, "I declare that the Church has no authority whatsoever to confer priestly ordination on women." (Women at the time, and since, responded that the Church was given that authority and exercised it in the early Church, since evidence of women priests has been found in the catacombs and other places. Also women were among the many first disciples beyond the twelve, and the twelve apostles were male to represent the new order of Jacob's twelve sons, the twelve tribes of Israel—among many other arguments.)

John Paul II admitted that the Church itself needed the biological and intellectual diversity of women in a deeper way for its own ecological and spiritual health—to reflect the Creator's and the Church's full representation. He also admitted that many of Jesus's most significant followers were women, to whom he revealed many of his key teachings and his Resurrection before he did to their male counterparts ("Letter to Women," 1995). The pope emphasized a number of times that "The Church is first and foremost the People of God, since all her members, men and women alike, *share*—each in his or her specific way—*in the prophetic, priestly and royal mission of Christ...*," and he repeated the concept that "*The universal priesthood of the faithful and the royal dignity* belong to both men and women." Then he addressed the question of power:

> Today in some quarters the fact that women cannot be ordained
> priests is being interpreted as a form of discrimination. But is this
> really the case? Certainly, the question could be put in these terms
> if the hierarchical priesthood granted a social position of privilege
> characterized by the exercise of "power." But this is not the case: the
> ministerial priesthood, in Christ's plan, is an expression not of *domina-
> tion* but of *service*! Anyone who interpreted it as "domination" would
> certainly be far from the intention of Christ, who in the Upper Room
> began the Last Supper by washing the feet of the Apostles.

Women, however, have struggled with his explanation, since certainly the religious sisters within the Church have been offering no less service, yet they have no say in how the Church as a whole is run. He was clearly torn. In 1996, John Paul wrote to the bishops about consecrated life and noted honestly:

> Certainly, the validity of many assertions relating to the position of
> women in different sectors of society and of the Church cannot be
> denied. It is equally important to point out that women's new self-
> awareness also helps men to reconsider their way of looking at things,

the way they understand themselves, where they place themselves in history and how they interpret it, and the way they organize social, political, economic, religious and ecclesial life.

Having received from Christ a message of liberation, the Church has the mission to proclaim this message prophetically....Consecrated women therefore rightly aspire to have their identity, ability, mission, and responsibility more clearly recognized, both in the awareness of the Church and in everyday life....

It is therefore urgently necessary to take certain concrete steps, beginning by providing room for women to participate in different fields and at all levels, including decision-making processes, above all in matters which concern women themselves....

In the field of theological, cultural and spiritual studies, much can be expected from the genius of women, not only in relation to specific aspects of feminine consecrated life, but also in understanding the faith in all its expressions.... In fact, "women occupy a place, in thought and action, that is unique and decisive. It depends on them to promote a new 'feminism' which rejects the temptation of imitating models of 'male domination,' in order to acknowledge and affirm the true genius of women in every aspect of the life of society, and overcome all discrimination, violence and exploitation." Here is reason to hope that a fuller acknowledgement of the mission of women will provide feminine consecrated life with a heightened awareness of its specific role and increased dedication to the cause of the Kingdom of God.

This will be expressed in many different works, such as involvement in evangelization, educational activities, participation in the formation of future priests and consecrated persons, animating Christian communities, giving spiritual support, and promoting the fundamental values of life and peace. To consecrated women and their extraordinary capacity for dedication, I once again express the gratitude and admiration of the whole Church, which supports them so that they will live their vocation fully and joyfully, and feel called to the great task of helping to educate the woman of today.

APOSTOLIC LETTER, *VITA CONSECRATA*, ON THE SOLEMNITY
OF THE ANNUNCIATION OF THE LORD, MARCH 25, 1996

As the pope aged and grew more ill and tired, the tone of his papal statements began to change and efforts shifted from looking for ways of broadening women's power and leadership to reining them in. In 2001, Catholic women and men, religious and lay, from 37 countries met in Dublin at the first Women's Ordination Worldwide (WOW) Conference to discuss the situation. (The Vatican warned some religious not to attend, such as Sister Joan Chitister, OSB. Those who did attend, including Chitister, relied on their informed consciences and prayer to make the final decision.) The conference passed eleven resolutions of action that included, among others:

- respectful calls to the pope to ask him to revoke the ban on *discussing* women's ordination, to restore women in the diaconate program, to make suitable theological training courses open to them, and to use inclusive language in all Church communications, sacraments, and liturgies, including the creeds;
- encouragement to participants and member organizations to pursue dialogues with their local bishops, religious, priests and laity on these requests and women's ordination; and
- establishment of an annual day of prayer on March 25 to invoke God's guidance and grace on women's ordination. (It has since been suggested that as many men in Catholic leadership don't notice how much women do for the Church, perhaps there should be a worldwide prayer strike where all women (and sympathetic men), including the "church ladies," just stay home from Mass the third weekend in March, praying together on their own to make their presence known by their absence.)

Pope John Paul II's response was not receptive, but rather a sharpening of the same cutting edges, and Pope Benedict XVI later launched an investigation into the orders of religious sisters and their obedience to the Church, chastising them for not thinking, speaking, and acting exactly like the male hierarchy in their interpretation of Scripture and Catholic doctrine.

Pope Francis Seeks Fruitful Service from All

Pope Francis started out with a somewhat similar approach in a meeting with eight hundred nuns in March 2013, telling them that they need to keep their integral, obedient connection to the Church and be nurturing spiritual mothers, like Mary, not "spinsters." (The tone was interpreted as negative slur to both the women religious and unmarried women.) In July he reiterated the decision of John Paul II's that women could not be made priests. He did, however, in an interview late that summer, call for a more inclusive Church, more innovative and welcoming to all:

> This Church with which we should be thinking is the home of all, not a small chapel that can hold only a small group of selected people. We must not reduce the bosom of the universal Church to a nest protecting our mediocrity...
>
> Instead of being just a Church that welcomes and receives by keeping the doors open, let us try also to be a Church that finds new roads, that is able to step outside itself and go to those who do not attend Mass, to those who have quit or are indifferent. The ones who quit sometimes do it for reasons that, if properly understood and assessed, can lead to a return. But that takes audacity and courage.
>
> "A BIG HEART OPEN TO GOD," *AMERICA* MAGAZINE, SEPTEMBER 30, 2013

Known for bringing fresh air into many areas of the Church to rebuild it, Francis has said that canonical law needs to be revamped, and he plans to do it. Perhaps, then, he will take Abigail Adams' counsel to her husband seriously: "Remember the ladies." Lay people are also saying, don't forget us!

In July 2013, Pope Francis established a powerful lay commission of seven financial, management, and communications experts from around the globe to reorganize the Vatican administration. The only clergy member is the secretary. This group includes one woman, and what a nontraditional choice she is—a stylish, thirty-something public relations specialist who is married, likes to tweet, and is extremely public and frank: Francesca Immacolata Chaouqui, from the Rome office of Ernst and Young. She has already supplied plenty of fodder for the rumor mills, ruffling the cassocks, as some say. Cardinals have been told they must keep no documents secret from this commission, and it reports only to the pope.

In addition, Francis set up a group of eight cardinals around the world to revise John Paul II's *Paster Bonus* constitution, updating and decentralizing it, delegating more authority and freedom in decision-making to regions and dioceses. Francis has expressed his desire to sweep away some vestiges of the medieval court system that has been carried along through the ages at the Vatican.

In October 2013, the pope told Eugenio Scalfari of the Italian newspaper *La Repubblica,* "The court is the leprosy of the papacy"—a system of flattery and narcissism within the Vatican linked with a great deal of temporal power and secrecy in money systems. This why he wants reform—to get away from Vatican-sustaining decisions and "clericalists." "This Vatican-centric view neglects the world around us," he told Scalfari, "I do not share this view and I'll do everything I can to change it."

God's guiding wisdom within the Church, he says, does not just come down from the top, but is part of the Holy Spirit working through the whole, believing people of God. Pope Francis explained that revelation and progress is messy and unclear while in its midst:

> And the Church is the people of God on the journey through history, with joys and sorrows. Thinking with the Church, therefore, is my way of being a part of this people. And all the faithful, considered as a whole, are infallible in matters of belief, and the people display this *infallibilitas in crededendo,* this infallibility in believing, though a supernatural sense of the faith of all the people walking together...
>
> When the dialogue among the people and the bishops and the pope goes down this road and is genuine, then it is assisted by the Holy Spirit. So this thinking with the Church does not concern theologians only...
>
> "A BIG HEART OPEN TO GOD," *AMERICA* MAGAZINE, SEPTEMBER 30, 2013

A Time of New Pastoral Forms
to Enhance and Enlarge the Old Ones

Pope Francis has certainly spurred conversations and longed-for hope among many who have been disappointed with the Church. In calling for dialogue, he acknowledges that new, more holistic ways of looking at the questions of "women's ordination" and "married priests" could be found within the larger scope of the Holy Spirit guiding the Church to its next stages of growth. Some people are calling for a Third Vatican Council to examine and throw out archaic canonical laws and authority structures that are not scriptural, but have been passed on from the medieval king-based model the pope dislikes. They urge the church to reinstate a leadership service model and some consensus decision making instead of a king-bishop model. They also long for more forms for sacramental ministry, ones based on the ordained charisms, or "chief orders" of the early Church (such as preachers, teachers, healers, miracle workers, speakers in tongues, deacons, administrators).

For the time being, Francis seems content to be searching, along with the Church and the women with in it. He told the Jesuit interviewer:

> I am wary of a solution that can be reduced to a kind of "female *machismo*" because a woman has a different makeup than a man.
>
> Women are asking deep questions that must be addressed. The Church cannot be herself without the woman and her role. The woman is essential for the Church. Mary, a woman, is more important than the bishops. I say this because we must not confuse the function with the dignity.
>
> We must therefore investigate further the role of women in the Church. We have to work harder to develop a profound theology of the woman. Only by making this step will it be possible to better reflect on their function within the Church. The feminine genius is needed wherever we make important decisions. The challenge today is this: to think about the specific place of women also in those places where the authority of the Church is exercised for various areas of the Church.
>
> "A BIG HEART OPEN TO GOD," *AMERICA* MAGAZINE, SEPTEMBER 30, 2013

Pope John Paul II had advised that women should have decision-making authority particularly over things pertaining to them. Thus, if there is to be a theology of women, shouldn't women develop it? And if women do, will they interpret it the same way the male Church hierarchy has?

Perhaps the time has come for a Pontifical Academy study week on the evolution of pastoral ministries in the early Church, with a special view to women, married people, and single lay people. Significant female scholars and archeologists should lead and be represented in higher or equal numbers to men, and lay people represented in higher numbers than clergy to present the scriptural, historical, and archeological evidence, and their recommendations for future forms of service

In Pope John Paul II's letter to bishops about consecrated life, he suggested that new formations of ministries can, will, and *should be* established creating new ways for women and all people within the Church to participate in decision-making levels that are as yet unformed. What is clear is that although John Paul shut the door on the priestly ordination of women, he couldn't really shut the door on women ministers and decision-makers in the church, for women were already standing in the Church's doorway with the door ajar, a position that he, in fact, arranged.

Some wonder what the Church would look like today if all of his ecological wisdom and counsel for the world had been applied to the microcosm of the Church itself. Perhaps more dioceses and parishes throughout the world would already be converted to alternative energies and sending service groups out to work on reforestation, river, and ocean issues. Many suggest that if women had had real shared power over decisions in the parishes and dioceses, the sexual abuse of children would never have been tolerated for any length of time, covered up, or swept under the rug. (Case in point: one of the Church's first female canon lawyers, Jennifer Hasleberger, resigned from her position as chancellor of canonical affairs in the Archdiocese of St. Paul and Minneapolis in October 2013 and became a whistleblower about diocesan cover-ups of clergy sexual abuse, forcing the archbishop to finally commit to making the list of abusing priests public.) Women have always been good at cleaning up and caring for their children. They've had millennia of experience.

No one will ever know. But we do know that as we give girls—all girls and women the world over—a chance for an education, microloans, freedom from violence, and access to leadership, as Pope John Paul II advised, they will be well on their way to doing their part to clean up our "beautiful endangered world."

The Elephant in the Room of Creation

Women, however, are not the only sexually defined class of people marginalized in the human family of creation and the Church. Scientific evidence shows that a fairly consistent one tenth of the population of humans (as well as other species) is homosexual, yet many people in this large percentage do not feel welcome in the Catholic Church. When Pope John Paul II was considering the concept of homosexuality, he stated:

> I wish merely to read what is said in the *Catechism of the Catholic Church*, which, after noting that homosexual acts are contrary to the natural law, then states: "The number of men and women who have deep-seated homosexual tendencies is not negligible. This inclination, which is objectively disordered, constitutes for most of them a trial. They must be accepted with respect, compassion and sensitivity. Every sign of unjust discrimination in their regard should be avoided.
>
> These persons are called to fulfill God's will in their lives and, if they are Christians, to unite to the sacrifice of the Lord's Cross the difficulties they may encounter from their condition" (CCC 2358).
>
> ANGELUS, JUBILEE IN PRISONS, JUNE 9, 2000

The *Cathecism* he quoted used the language of professional psychology's *Diagnostic and Statistical Manual (DSM) II* from the 1950s and 1960s, which listed homosexuality as one of the sexual deviation "disorders." However, this listing was dropped in the *DSM III* and beyond. Psychologists (and the Church) make distinctions between homosexuality forced upon a person by circumstances (such as traumatic incidents or child sexual abuse) versus inherent orientation. Recent scientific studies suggest that homosexuality is caused by certain "epi-marks" on specific genes that determine how they are activated in terms of sexual orientation. Complicated stuff, of course, all happening when these children were knit in their mother's wombs, for

God's own reasons. Scripture proclaims that God has blessed all of creation, and every element of it, as "good."

The science is complex enough to suggest that it may be worth the Pontifical Academy of Sciences doing a thorough examination of this. Catholics, of course, would not want to be labeling something as *against natural law* or inclinations as *disordered* that may be part of God's planned diversity in all species.

The U.S. Conference of Catholic Bishops has a compassionate pastoral statement, "Always Our Children," based on Pope John Paul II's more extensive teachings on sexuality and the *Cathecism*. It states:

> There seems to be no single cause of a homosexual orientation. A common opinion of experts is that there are multiple factors—genetic, hormonal, psychological—that may give rise to it. Generally, homosexual orientation is experienced as a given, not as something freely chosen. By itself, therefore, a homosexual orientation cannot be considered sinful, for morality presumes the freedom to choose.

The pastoral letter urges parents not to reject their children for their innate makeup but to love and support them all the more. It also states that one shouldn't try to deny or change the person but guide them toward lifelong abstinence, because sexual acts outside of marriage are considered immoral, and sacramental marriage for homosexuals is barred by the Church.

The unintended consequence of this theological position has been to set aside gay youth and adults as negatively different in society and the Church: that since marriage is barred, any expression of their normal sexuality and tendencies is wrong, considered immoral, and shameful. Many feel isolated and discriminated against in society and in the Church. Some feel alienated from God and simply leave the Church…others go in the opposite direction and enter consecrated religious life, where their welcome can also be ambiguous.

Too many homosexual individuals, especially youth, are bullied, badgered, and shamed—*some to the point of suicide.* Young people of conscience see this and ask, Why? How does this all fit with loving all the family of God's creation? To many it seems like just another form of racism, one not based on color.

The youth also witness the impassioned and oftentimes mean-spirited political lobbying by Catholics and Christians against gay marriage, *but not against climate change*, and they ask, Why? (As a mother of teens and a confirmation teacher, I know.) If a legal marriage is different from a Church marriage, what's the big deal? Isn't that the separation of Church and State?

They also hear that the Old Testament outlawed homosexual acts, but weren't women considered property at the time, with men having as many wives and slave concubines as possible to beget as many children as they could? Obviously homosexuality isn't going to work to grow a nation. And if a man had to marry his brother's widow so even a dead guy could have heirs, are men still supposed to do that too? If not, why are the rules about homosexuality still kept? They also challenge: What did Jesus say about it?

Of course, the answer is nothing. He talked about divorce, but not homosexuality. All right, they say, if it wasn't a priority for him, why is it for us?

The youth even go a step further. Isn't committed love for gay people better than having them hate themselves for who they are, commit suicide, or sleep around? And if God made them to desire people of the same sex, why would He do that and then tell them they're wrong to feel that way? Many say that if all this furor over homosexuality is God talking, then God sounds pretty "lame," and so does Catholicism and Christianity in general.

This is a human ecological issue that affects society's ability to work together. Can we afford not to listen to the legitimate questions of our youth? And of 10% of God's people? Does not Christ need them, and all of us need them and their energies and ideas to face the challenges in the Church and on this planet?

Pope Francis seems to have heard the questions and concerns of the youth, for on a plane back from Brazil, Francis said to a reporter in a press conference "If someone is gay and he searches for the Lord and has good will, who am I to judge?" (JULY 29, 2013).

Created or "Disordered" Sexual Orientation?

Alan Chambers spent years running a Christian ministry called Exodus, trying to convert homosexuals away from their "disordered" inclinations. In spring 2013, Chambers quit and came out with a deeply felt public apology:

> Please know that I am deeply sorry. I am sorry for the pain and hurt many of you have experienced. I am sorry that some of you spent years working through the shame and guilt you felt when your attractions didn't change....I am sorry that there were times I didn't stand up to people publicly "on my side" who called you names like sodomite—or worse....
>
> More than anything, I am sorry that so many have interpreted this *religious* rejection by Christians as God's rejection. I am profoundly sorry that many have walked away from their faith and that some have chosen to end their lives....
>
> I cannot apologize for my deeply held biblical beliefs about the boundaries I see in scripture surrounding sex, but I will exercise my beliefs with great care and respect for those who do not share them. I cannot apologize for my beliefs about marriage. But I do not have any desire to fight you on your beliefs or the rights that you seek. My beliefs about these things will never again interfere with God's command to love my neighbor as I love myself.
>
> ALAN CHAMBERS, AS REPORTED IN *CHRISTIANITY TODAY*, JUNE 19, 2013,

Chambers certainly offers perspectives to ponder. When asked about his comment about not judging people who are gay, Pope Francis explained: "I said what the *Catechism* says. Religion has the right to express its opinion in the service of the people, but God in creation has set us free: it is not possible to interfere spiritually in the life of a person."

Francis had earlier pointed out that the Catholic Church can easily fall prey to the arrogance of being self-referential for truth, and if it is, it can miss truths that God is trying to communicate in the present and future. It ends up pointing to itself instead of to Christ. "All this demands that we keep our heart and mind open, avoiding the spiritual illness of self-referentiality. When the Church becomes self-referential she too falls ill and ages. May our gaze, firmly fixed on Christ, be prophetic and dynamic in looking to the future."

(ADDRESS TO THE COMMUNITY OF WRITERS OF LA CIVILTA CATTOLICA, JUNE 14, 2013).

In August of 2013, having been a pope six months, Francis engaged in an in-depth interview with a fellow Jesuit in Italy, which was reprinted in *America*. He meditated in more depth about how the Church needs to refocus itself on core issues:

> We cannot insist only on issues related to abortion, gay marriage and the use of contraceptive methods.... It is not necessary to talk about these issues all the time. The dogmatic and moral teachings of the church are not all equivalent. The Church's pastoral ministry cannot be obsessed with the transmission of a disjointed multitude of doctrines to be imposed insistently.
>
> Proclamation...focuses on the essentials, on the necessary things...what makes the heart burn, as it did for the disciples at Emmaus. We have to find a new balance; otherwise even the moral edifice of the Church is likely to fall like a house of cards, losing the freshness and fragrance of the Gospel.

"A BIG HEART OPEN TO GOD," *AMERICA* MAGAZINE, SEPTEMBER 30, 2013

On the main, the pope wants the Vatican and Church as a whole to rethink who their roles in serving Christ and what it needs to focus on to help heal the world:

> I see clearly that the thing the Church needs most today is the ability to heal wounds and to warm the hearts of the faithful; it needs nearness, proximity. I see the Church as a field hospital after battle.... The Church sometimes has locked itself up in small things, in small-minded rules.

"A BIG HEART OPEN TO GOD," *AMERICA* MAGAZINE, SEPTEMBER 30, 2013

To be this field hospital and minister to those hurting and accomplish all we need to save our planet and life as we know it, the Church will have to be able to engage everyone's energies, talents, and creativity: women, men, and youth, heterosexual and homosexual, Christian and believers of other faiths, people religious and secular, black, white, red, yellow, and brown—and green. Pope John Paul II told us:

> We must pull down the walls of division, hostility, and hate so that the family of God's children may once again live in harmony at the one table, to bless and praise the Creator for the gifts he lavishes upon all without distinction (cf. Mt. 5: 43–48).

GENERAL AUDIENCE, NOVEMBER 17, 2004

Population growth must be faced not only with the exercise of responsible parenthood, which respects divine law, but also by economic means, which have a profound effect on social institutions.

ADDRESS TO THE PONTIFICAL ACADEMY OF SCIENCES, STUDY WEEK ON
RESOURCES AND POPULATION, NOVEMBER 22, 1991

7. Population Growth, Limited Resources, and Responsible Parenthood

The world population is presently at over 7 billion, up from 3 billion in 1960. The U. S. Census Bureau's Population Clock gives continual tabulations; the present rate of increase is: "One birth every 8 seconds; one death every 12 seconds; one international migrant (net) every 44 seconds; net gain one person every 13 seconds." At this rate, by 2040, counting disasters and epidemics, we'll be at nearly 9 billion. Nine billion people who need food and shelter, who produce waste, and who need social services such as education, health care, and, of course, water.

Then you must factor in the globalization of consumerism and consumption that is shrinking wildlife habitats, worsening pollution, intensifying droughts, and depleting fresh water resources. Suddenly it becomes obvious that more people leads to infinitely multiplying ecological crises, moral crises, and strains on natural resources all over the world.

Psychologically, too, there is an enormous toll. Bangladesh, for instance, a nation slightly smaller than Iowa, has half the population of the entire United States, with a fraction of its resources. It is poor and crowded, and it's not unusual to see beggars who have actually disfigured themselves to solicit sympathy and handouts to survive. Those who can, emigrate to other countries with more space and opportunity.

Rich countries feel the psychological effects, too. They argue about immigration. In America, people talk about how we used to be much more laid back, with less pressure to be working constantly. That's certainly because contemporary technology allows people to be on call all the time, but it's also because the greater the population, the greater the competition for jobs and resources, which adds to stress, tension, and aggression, as it would in any overcrowded chicken pen. Because of this, population control is not the concern of one nation or region of the world, but of all areas, for all are ultimately connected and affected.

Recognizing this, Pope John Paul II met the question head on and sponsored a Pontifical Academy of Sciences Study Week on Resources and Population in 1991. He addressed the members and the world:

> I thank you for having accepted the invitation of the Pontifical Academy of Sciences to take part in a scientific discussion...of great concern to society today: *the relationship between the accelerated increase in world population and the availability of natural resources....* The Church is aware of the complexity of the problem. It is one that must be faced without delay....
>
> We all have precise duties towards future generations: this is an essential dimension of the problem, and it impels us to base our proposals on solid prospects regarding population growth and the availability of resources.

> ADDRESS TO THE PONTIFICAL ACADEMY OF SCIENCES, STUDY WEEK ON RESOURCES AND POPULATION, NOVEMBER 22, 1991

His ensuing teachings addressed the wide-ranging complexities of population issues and their interrelated physical, demographic, economic, and spiritual aspects.

Urgent Population Consideration

Contrary to what many have thought, Pope John Paul II acknowledged the immediacy of the situation and the urgent need for action, which few people within or outside the Catholic Church really understood. He *affirmed the need for governments to establish population ideals and guidance,* but he reminded them to integrate a respect for human freedom in their application and enforcement.

It is the responsibility of the public authorities, within the limits of their legitimate competence, to issue directives that reconcile the containments of births with respect for the free and personal assumption of responsibility by individuals.

ADDRESS TO THE PONTIFICAL ACADEMY OF SCIENCES, STUDY
WEEK ON RESOURCES AND POPULATION, NOVEMBER 22, 1991

He advocated a multipronged approach to population planning, policies, education, and procreation, based on eight ethical principles:

1. Understand that different regions have different situations and issues.
2. Respect the freedom and dignity of families and their desire for children.
3. Encourage responsible parenthood.
4. Value all children and provide community support as needed.
5. Offer economic and educational opportunities, especially to women, to encourage smaller families and responsible nonintrusive family planning.
6. Share—those individuals and countries with more must live more simply and offer resources to those with less.
7. Consider future generations.
8. Work in solidarity for sustainable development.

Different Regions, Different Situations

Having traveled so extensively, the pope understood that a one-size-fits-all approach would *not* work. He warned against a new form of imperialism, with richer countries telling less developed countries what to do about population.

The close connection between the world's resources and its inhabitants must be evaluated, as you have opportunely done, by also taking into account the present imbalances in demographic distribution, in movements of migrants, in the allocation and distribution of resources....

ADDRESS TO THE PONTIFICAL ACADEMY OF SCIENCES, STUDY
WEEK ON RESOURCES AND POPULATION, NOVEMBER 22, 1991

Yet he was always emphatic that to be both moral and practically successful, population policies could not work against people's God-given natural desire and right to have children and build family units, and for couples to
make their own decisions. Thus he decried laws such as the one-child rule or
others that promote or turn a blind eye to abortion, infanticide, or steriliza-
tion as family planning or birth control methods:

> Human society is first and foremost a society of persons, *whose inalien-*
> *able rights must always be respected.* No political authority, whether
> national or international, can ever propose, much less impose, a policy
> that is contrary to the good of persons and of families....
>
> The urgency of the situation must not lead into error in propos-
> ing ways of intervening. To apply methods that are not in accord with
> the true nature of humans actually ends up by causing tragic harm. For
> this reason, the Church...upholds the principle of responsible parent-
> hood and considers it her chief duty to draw urgent attention to the
> morality of the methods employed. These must always respect the per-
> son and the person's inalienable rights.
>
> ADDRESS TO THE PONTIFICAL ACADEMY OF SCIENCES, STUDY
> WEEK ON RESOURCES AND POPULATION, NOVEMBER 22, 1991

Encourage Responsible Parenthood and Respectful Restraint

Pope John Paul II urged people to use their sexual and regenerative abilities prudently and respectfully, with each family and all individuals balanc-
ing their freedom to have children with their responsibilities to care for
them *with limited resources*, those of their household and the world. He did
not shame or deny the longstanding love of the Catholic Church for large
families, but put the issue into a larger context—the family of humanity.
He asked the Church, public authorities, and all parents to educate their
consciences in this broader context and to *exercise restraint*. Even before the
study week, he had acknowledged the problematic nature of the population
situation and was preaching respectful limits to families, as in a homily he
gave in Burundi in 1990: "As for the population problem, *the primary respon-*
sibility falls naturally on the parents..." Informed by the Academy's findings,

he became more vocal about each family's moral and faith responsibility to look beyond their own needs, wants, and access to resources, to encompass the wider picture of their effect on the world at large:

> It is a task first of all for the family, the basic unit of society. The family draws moral strength from parents' sense of responsibility...which includes a balanced attitude toward procreation, an attitude that seeks to build a more united and caring society.
>
> ADDRESS TO THE PONTIFICAL ACADEMY OF SCIENCES, STUDY
> WEEK ON RESOURCES AND POPULATION, NOVEMBER 22, 1991

As Bill McKibben emphasized in his book *Enough,* humans are uniquely gifted in their capacity to set and respect limits, to exercise reasoned restraint, and to apply self- and societal discipline when they choose. Self-control and discipline have long been heralded values in religious and communal societies, but they have lost favor in contemporary Western culture. John Paul II, however, tried to bring back the freeing aspects of limits, explaining how loving self-discipline and sacrifice are enduring tools for living a healthy, whole, and holy life: "People need to rediscover the moral significance of respecting limits; they must grow and mature in the sense of responsibility with regard to every aspect of life" (STUDY WEEK ON RESOURCES AND POPULATION).

Natural Family Planning, Accessible to All at Little Cost, No Violence

Pope John Paul II naturally promoted various methods of Natural Family Planning (NFP) as the ideal, family-based approach for limiting population. Unfortunately, most people think of the famous and faulty "rhythm method," the one-size-fits-all method of tracking the days of a menstrual cycle and having sex only when the woman was predicted by the count to be at a low fertility point. Many jokes have been made of its low success rate; it was very trying for the women and men who had been told to use it when it didn't work consistently.

Thankfully, science has offered far more information to women about how to read the signs their bodies give them about their fertility, such as the calendar method, the symptothermal method (measuring basal body temperature), and the ovulation method (observing cervical mucous) to naturally space children. There is also the more expensive Marquette Method, which uses a battery-operated ClearBlue Fertility Monitor and urine strips to check estrogen levels and other indicators to predict ovulation.

The pope recommended NFP methods for many reasons: they are based on natural, noninvasive processes; they encourage women to better know their own bodies and natural cycles; they are free (except for the cost of a thermometer, or monitor and strips if desired), and thus are available to those of little means if they are taught how to implement them; they encourage communication and respect between partners. In ideal situations, they give women the power over sexual interaction and the man the chance to honor her. They are not a quick fix that can encourage irresponsibility or lack of respect; and they can be used to get pregnant as well as not. So John Paul II advocated that governments and nonprofits invest in free NFP training in cities and villages and through faith communities:

> Using the *natural methods* requires and strengthens the harmony of the married couple, it helps and confirms the rediscovery of the marvelous gift of parenthood, it involves respect for nature and demands the responsibility of the individuals. According to many authoritative opinions, they also foster more completely that human ecology which is the harmony between the demands of nature and personal behavior.
>
> At the global level, this choice supports the process of freedom and emancipation of women and peoples from unjust family planning programs which bring in their sad wake the various forms of contraception, abortion, and sterilization.
>
> "TO THE TEACHERS OF NFP," *L'OSSERVATORE ROMANO*, JANUARY 22, 1997

It works best, of course, when both partners attend training and work on it together. Women who use any NFP method will attest that there are some drawbacks, including putting a great deal of the burden on them to constantly be checking their signs and having to be the timekeeper and referee, so to speak. It also means that if a couple is trying to space their children out or to avoid a pregnancy, the woman will not be having sex when her body most feels like having it, and will probably be having it when she doesn't. In addition, fertility signs are not always consistent and clear. These are messy, small details that the pope, cardinals, and priests would not really understand firsthand, but they can be dealt with, as in any contraceptive approach. The bottom line is that in many situations NFP can strengthen a marriage by bringing a couple closer in planning and intimacy, with cuddling, conversation, and praying together as important as making love.

The Unaddressed Difficulties of the Social Realities

However, not all situations are ideally suited for Natural Family Planning. One essential consideration that *Humanae Vitae* (Pope Paul VI's Church document on birth control) and subsequent pope and Church positions did *not* adequately address is what women can or should do in situations of violence, when they are in unequal and threatening power dynamics and don't have the choice over the sexual interaction—which we now know is the situation for a high percentage of the world's women. Or what they should do when they are unable to get their husbands to follow their lead on timing, or stop having affairs that can bring sexual diseases? There are so many women who have no access to training about their bodily rhythms and signals, or even the watches, clocks, and alarms to make sure they take their temperature at the exact same time each day. What about those who are receiving chemical or radioactive treatments for cancer or other illnesses, whose cycles are thrown off? Then there are those couples who have very limited, sporadic time together—what if those times don't coincide with her cycle? There are so many situations in which the NFP approach becomes the opposite of the ideal. In these circumstances, as in all others, Catholic teachings say that one needs to make the best moral choice using an informed conscience and the resources at hand.

However, with so very many ethical exceptions and moral land mines because of the real-life messiness of marriage and relationships, many Catholics wonder if aspects of *Humanae Vitae* need to updated with some female and lay input. A 2011 Guttmacher Institute study of sexual behavior found that only 2% of sexually active American Catholic women of childbearing age (including committed churchgoers), who want to avoid a birth use NFP. Though the study was flawed in segmenting out only the people most likely to have used contraception, it did indicate that a large percentage do not use NFP. Studies range, but somewhere between 75% and 98% of Catholics globally don't agree with the Church's recommendation. To many, it seems arbitrary and unreasonable, not based on Christ or Scripture or core beliefs. Another study shows 81% of American Catholics think that married couples should use their own consciences to decide how to plan the growth of their family.

In general, many feel that neither John Paul II nor the Church as an institution has yet adequately considered the painful realities of the status of women in many Catholic cultures, and the preponderance of male infidelity and contagious sexual diseases, or the full dynamics of intimate relationships.

Push for a More Moderate, Flexible Stance

It is clear that over the centuries the Church's strong procreation positions have had a drastic effect on increasing the world's population. And this is no small concern. Pope John Paul II declared that the population situation was more than a planning challenge—it was a *pressing moral problem that had to be addressed not in the future, but now.* In his 1999 apostolic letter to the bishops of Asia, he wrote, "Several Asian countries face difficulties related to population growth, which is 'not merely a demographic or economic problem but especially a moral one.'" Thus, responsible limits and methods must be sought as matters of faith to care for families and all of creation.

Priests, nuns, and faith-filled Catholic lay people in and out of the medical fields have implored the popes, the Vatican, and the bishops to revise the Church's stance on contraception, particularly the ban on condoms. Uganda, for instance, has seen dramatic reductions in its high HIV/Aids rate through a campaign promoting abstinence outside of marriage, monogamy, and condoms. None of the three alone seems to have the same effectiveness, so Catholic advocates see condoms as an essential public health and population reduction tool. Advocates point out that condoms are not evil in and of themselves but how and why they are used. Within a marriage, condoms can be utilized in ways that affirm all life values and interpersonal communication. By preventing many unplanned and unprepared-for births where

NFP doesn't ideally fit, condoms can help stop hunger and malnutrition in the millions of babies born to financially overstressed families, reducing the temptation for abortion and child abuse.

Education for Women: The Most Effective Natural Family Planning Approach

Understanding why people have children is key to finding the most effective approaches. People in poverty-stricken countries often have large families because infant and child mortality rates are high, and parents need to ensure that a certain number of their children grow up to take care of them when they grow old. In some cultures, girls are married at very young ages and they don't know how to prevent pregnancies, or if they morally should (and some end up dying in childbirth). The financial strain of too many children and the health strain on the mothers of carrying, bearing, and nursing numerous infants in a row make it hard for these children to end up healthy, well nourished, and educated. The cycle of poverty then deepens, for the family and the country.

Because of this, Pope John Paul II reminded governments that "population growth has to be faced not only by the exercise of a responsible parenthood…but also by economic means that have a profound effect on social institutions" (STUDY WEEK ON RESOURCES AND POPULATION). He advised nations to look realistically at their particular population-resource difficulties and projections in the context of their neighbor nations and the world as a whole. They must also consider the effects of immigration and refugees. Key problems to be addressed are distribution of food, housing, jobs with decent wages, clean water, air, and environment, and room for wildlife.

Governments can then set realistic but flexible goals for births and family sizes and immigration. While maintaining that all family choices must be voluntary, governments can promote family size targets through cultural awareness programs, public-health resources, and education. But they will also have to find ways to redistribute access to economic opportunities out of the hands of the few and into the hands of the many, to make sure all people have access to dignified ways to support themselves:

A political program that respects the nature of the human person can influence demographic developments, but it should be accompanied by a redistribution of economic resources among the citizens. Otherwise, such provisions can risk placing the heaviest burden on the poorest and weakest sectors of society, thus adding injustice to injustice.

ADDRESS TO THE PONTIFICAL ACADEMY OF SCIENCES, STUDY
WEEK ON RESOURCES AND POPULATION, NOVEMBER 22, 1991

Education and professional training are key, as they open doors to opportunities and teach new modes of thinking to broaden perspectives. John Paul II stated:

A strong common commitment to institutional reform is needed, a commitment that aims at raising the level of intellectual and personal maturity by means of a satisfactory educational system.

It will also aim at strengthening enterprise and the creation of jobs through adequate investments....

Particularly in the developing countries, where young people represent a high percentage of the population, it is necessary to eliminate the grave shortage of structures for ensuring education, the spread of culture, and professional training.

ADDRESS TO THE PONTIFICAL ACADEMY OF SCIENCES, STUDY
WEEK ON RESOURCES AND POPULATION, NOVEMBER 22, 1991

Studies have borne out the pope's wisdom. Educating girls and women so that they are empowered and have more choices about their lives has proven *the most effective of all approaches to population modulation.* When women (and men) are aware of options and have opportunities, they tend to have fewer children.

Those with Much Must Live More Simply, and Offer Resources to Those with Less

Pope John Paul II also noted that the population control efforts traditionally urged on poor countries by richer ones have often masked the overuse of the developing countries resources by the developed nations, draining options, possibilities, and hope. Resource depletion arises not only from large populations using a little but from smaller, more affluent populations each using much more, through overconsumption and pollution. John Paul explained that, "often it is precisely the latter countries [with dwindling, aging populations], with their high levels of consumption, that are most responsible for pollution to the environment" (STUDY WEEK ON RESOURCES AND POPULATION).

Do not be guided simply by fear of shortages, he told national and international leaders, but engage in a careful evaluation of the diverse and multileveled aspects of the problem:

> Various aspects of life in society are involved here, from family rights to regulation of land ownership, from social welfare to the organization of labor, from public order to ways of establishing a consensus in society.

The only way to turn the situation around would be a radical readjustment of priorities by those who revel in abundance. Individuals, businesses, and governments in developed countries have to examine their patterns of use and then strive to use and waste less, utilize more reusable products and renewable energies, clean up and protect resources, pay more for items manufactured by workers in developing countries, and vote for policies of more equitable distribution of goods and trade. It is morally unacceptable to have some people hunger while others waste because of unequal power situations and endemic injustice.

> Damage to the environment and the increasing scarcity of natural resources are often the result of human errors. Despite the fact that the world produces enough food for everyone, hundreds of millions of

people are suffering from hunger while elsewhere enormous quantities
of food go to waste.

ADDRESS TO THE PONTIFICAL ACADEMY OF SCIENCES, STUDY
WEEK ON RESOURCES AND POPULATION, NOVEMBER 22, 1991

Look to the Children with Joy and Support

The heart of the question for Pope John Paul II was that all conceived children *are wanted by God because of their very existence* and therefore need to be
well cared for *by us as the gifts they are*. Children bring joy and should never
be mere statistics on population charts. All children deserve care and need
the support not only of their family, he said, but also of their communities
and the world: "By the mere fact of being conceived, a child is entitled to
rights and deserving of care and attention; and someone has the duty to
provide these" (WORLD DAY OF PEACE 2005 MESSAGE, "DO NOT BE OVERCOME BY EVIL, BUT
OVERCOME EVIL WITH GOOD").

While the pope urged parents to think ahead and have children according to responsible limits, he also understood that things do not always go
as people "plan," and they must still embrace joyfully, or at the very least
with love and respect, the lives begun outside of planning or ideal situations. Morally, societies must be willing to help support the child with
access to water, food, shelter, clothing, safety and security, health services,
and education:

> It is true that a child represents the joy not only of its parents but also
> the joy of…the whole of society. But it is also true that in our days,
> unfortunately, many children in different parts of the world are suf-
> fering and being threatened: they are hungry and poor, they are dying
> from diseases and malnutrition, they are the victims of war, they are
> abandoned by their parents and condemned to remain without a
> home, without the warmth of a family of their own, they suffer many
> forms of violence and arrogance from grown-ups. How can we not
> care, when we see the suffering of so many children, especially when
> this suffering is in some way caused by grown-ups?

> "LETTER TO CHILDREN," YEAR OF THE FAMILY, DECEMBER 12, 1994

Having laid out his guiding principles, Pope John Paul II asked humanity to look with hope toward a healthy future, creating long-lasting population solutions and more responsible parenthood. It would not be fair for the generations of today to have large families leading to forced restrictions on later generations. He reminded people:

> As a member of the human family, each person becomes as it were a citizen of the world, with consequent duties and rights, since all human beings are united by a common origin and the same supreme destiny.
>
> WORLD DAY OF PEACE 2005 MESSAGE, "DO NOT BE OVERCOME BY EVIL, BUT OVERCOME EVIL WITH GOOD," JANUARY 1, 2005

John Paul hoped that people at every level of decision-making would consider their options and make choices to support a bright future for the children of tomorrow. In his final World Day of Peace address, he counseled:

> The appeal to each individual's sense of responsibility is an urgent one. So is the appeal for solidarity on the part of everyone.
>
> The dynamics of population growth, the complexity of uncovering and distributing resources, and their mutual independencies and environmental consequences constitute a long-term and demanding challenge.
>
> It is only through a more austere manner of living, one that springs from respect for the dignity of the person, that humanity will be able to meet this challenge adequately.
>
> In short, a renewed way of life is needed, one spread by authentic humanism and therefore capable of dissuading public authorities from proposing and legalizing solutions contrary to the true and lasting common good. *It is a manner of living that . . . will help to bring about a world in which love for others is accepted as the general and normative rule.*
>
> ADDRESS TO THE PONTIFICAL ACADEMY OF SCIENCES, STUDY WEEK ON RESOURCES AND POPULATION, NOVEMBER 22, 1991

*"You are stewards of some of the most important resources
God has given to the world.
Therefore conserve the land well,
so that your children's children
and generations after them will inherit
an even richer land than was entrusted to you."*

LIVING HISTORY FARMS, HOMILY, DES MOINES, IOWA, OCTOBER 4, 1979

8. Farming and Conservation

Tom Frantzen had been farming in New Hampton, Iowa, on the lands he had purchased from his father. It was 1979, and he was listening to Pope John Paul II's homily at the Mass held outdoors at the Living History Farm in Des Moines when Tom heard the pope say: "…conserve the land well, so that your children's children and generations after them will inherit an even richer land than was entrusted to you." Frantzen remembers, "I had been farming conventionally for about five years, using liquid manure, pesticides and herbicides, but that all changed when the pope visited Iowa and I heard him speak" (ORGANICVALLEY.COOP PROFILE, "TOM AND IRENE FRANTZEN RECEIVE SUSTAINABLE AGRICULTURE ACHIEVEMENT AWARD," JANUARY 5, 2011).

To the crowd of fifty thousand and those listening on the radio, Pope John Paul II told the farmers that they had a great responsibility in their hands and God was calling them to a threefold, faith-filled response: gratitude for the miracle of life in seeds and the earth; stewardship and conservation of this land and its life-giving properties; and generosity—sharing the fruit of the land and being part of a distribution system that does. He called farmers to align their hearts and minds with God the Creator and then act in his stead by caring for the lands they were given as a gift:

> …the land must be conserved with care since it is intended to be
> fruitful for generation upon generation. You who live in the heartland
> of America have been entrusted with some of the earth's best land: the

soil so rich in minerals, the climate so favorable for producing bountiful crops, with fresh water and unpolluted air available all around you. You are stewards of some of the most important resources God has given to the world.

...But also remember what the heart of your vocation is. While it is true here that farming today provides an economic livelihood for the farmer, still it will always be more than an enterprise of profit-making. In farming, you cooperate with the Creator in the very sustenance of life on earth.

The pope also reminded the crowd of the connection of their work to the very simple life and work of Christ:

In the life of Jesus, we see a real closeness to the land. In his teaching, he referred to the "birds of the air," the "lilies of the field." He talked about the farmer who went out to sow the seed; and he referred to his heavenly father as the "vinedresser," and to himself as the "good shepherd." This closeness to nature, this spontaneous awareness of creation as a gift from God, as well as the blessing of a close-knit family—characteristics of farm life in every age including our own—these were part of the life of Jesus. Therefore I invite you to let your attitudes always be the same as those of Christ Jesus.

Tom was struck right through to his heart. "I realized at that moment that I could not continue to farm the way I'd been farming. I knew that I had to be a better steward of the land and do what I could to preserve it for future generations" (ORGANIC VALLEY PROFILE, JANUARY 5, 2011). Tom turned to his wife, Irene, and together they committed themselves to change the way they farmed. They embraced their larger vocation of conserving the land and water while producing food, always with an eye to the future.

Transforming Priorities and Ourselves

A little more than thirty years later, in 2011, the Practical Farmers of Iowa presented Tom and Irene Frantzen with their Sustainable Agriculture

Achievement Award. This honor goes annually to a farmer, or farm family, who has been successful in a creating a biologically diverse, sustainable farm that produces healthful food, and who has contributed to the building up of a vibrant community. The Frantzens had participated in more than fifty research trials, set up an organic meat company, and taught other farmers how to attain sustainability.

Their transition, of course, was not easy or immediate. The Frantzens became the organic guinea pigs in their area, the first participants in their county's conservation stewardship program. They sought training in Holistic Resource Management from the Land Stewardship Project to write a five-year business plan to put the principles into practice on their 300 acres (and fifty rented acres). Crop rotation, natural fertilizer, and pest management techniques, along with care for the soil ecology, evolved as part of their everyday methods. Besides their organic hogs, cattle, corn, soybeans, succotash, and hay, they raised three children on this land, who actively helped with the chores and decisions. The youth created their own field experiments, 4-H, and FFA (Future Farmers of America) projects. "We have had, and continue to have, a good life on the farm," says Irene. "The land has been good to us and so we are good to the land" (ORGANIC VALLEY PROFILE, JANUARY 5, 2011.).

No chemical pollution runs from their lands into the rivers or wetlands, or seeps into the aquifers. No toxins reside in the produce and meat they sell. Their children grew up in a setting of natural beauty, faith, community engagement, and meaningful work—*a culture of life*. On a family farm in Iowa, they proved that everyone benefits when relationships are aligned.

These are actions of faith and God together in creation. Pope John Paul II told farm workers in California:

> The growth of the wheat and its maturing, which greatly depends on the fertility of the soil, comes from the nature and vitality of creation itself. Consequently there is *another source of growth*: the one who is *above nature and above the man who cultivates the earth.*
>
> In a sense, the Creator "*hides himself*" in this life-giving process of nature. It is the human person, with the help of intellect and faith,

who is called to "discover" and "unveil" the presence of God and his action in all of creation...

HOMILY, MASS FOR RURAL WORKERS IN LAGUNA SECA,
MONTEREY PENINSULA, CALIFORNIA, SEPTEMBER 17, 1987

The Damages of Industrial Farming

Throughout the world, there are as many different kinds of farms as there are species being raised, and these farms are not all easy on the earth. They fall into a few major categories based on their approach to the land: large-scale industrial or factory-style farms owned by corporations and run by management teams that hire farm workers; smaller family-run chemically supported farms; industrial organic farms; family-run organic or sustainable farms; subsistence farms; and finally, restoration permaculture farms. Each type faces different challenges ecologically and economically.

In the industrial farm, efficiency and output are the primary concerns, and the land, animals, and farm workers are viewed and treated whichever way best suits these financial goals. To accomplish production quotas, the operations use chemical pesticides, fertilizers, synthesized hormones, and genetically modified seeds. Animals are crammed into overcrowded spaces or cages, resulting in rampant disease, extreme suffering, and death. The surrounding areas are polluted by noxious and toxic odors; waters are polluted; soil ecology destroyed; wildlife chased off or poisoned by chemicals; pollinators and amphibians are under threat of deaths or stress from chemicals; and generally the farm workers (often migrants) are not paid well or offered particularly safe working conditions. These operations are productive on one level but not sustainable for the long haul without constant chemical additives. Wild forms of nature are eliminated on that land, for efficiency's sake. This is farming as a mere profit-oriented business, not a vocation in God's service.

The family farm versions of conventional farming are more personal and less drastic, but they share some of the same problems, as described in Michael Pollan's fascinating and seminal book *The Omnivore's Dilemma*. One is loss of topsoil. Topsoil is blown and washed away in conventional farming due to repetitive plowing at unsustainable rates—5.2 tons per acre in Iowa, for example, 10 times more than the rate the earth can rebuild it. Topsoil loss equals loss of fertility and a step toward desertification. The topsoil is also washed into rivers, causing sedimentation, ruining fish and aquatic habitats.

Ecologist Aldo Leopold called soil "the basic natural resource." He said:

> Destruction of the soil is the most fundamental kind of economic loss which the human race can suffer. With enough time and money, a neglected farm can be put on its feet—if the soil is there. By expensive replanting and with a generation or two of waiting, a ruined forest can be made productive—if the soil is there...But if the soil is gone, the loss in absolute and irrevocable.
>
> "EROSION AND PROSPERITY" SPEECH AT UNIVERSITY OF ARIZONA, 1921

Leopold also wrote, "The loss of our existing farms we dismiss as an act of God...On the contrary, it is the direct result of our own misuse of the country we are trying to improve" ("PIONEERS AND GULLIES," *THE RIVER OF THE MOTHER OF GOD*, 1924).

Pope John Paul II did not accept people laying the blame for loss of the soil at God's feet as a "natural disaster" either. At the Jubilee of the Agricultural World, he reminded all those involved in agriculture what their main mission should be:

> Without doubt, the most important value at stake when we look at the earth and at those who work, is the principle that brings the earth back to her Creator: *the earth belongs to God!* It must therefore be treated according to his law. If, with regard to natural resources, especially under the pressure of industrialization, an irresponsible culture of "dominion" has been reinforced with devastating ecological consequences, this certainly does not correspond to God's plan.
>
> JUBILEE OF THE AGRICULTURAL WORLD ADDRESS, VATICAN CITY, NOVEMBER 11, 2000

John Paul taught that the fertility of the land comes from the soil itself, for which we must be grateful and careful, for as goes the soil, so does abundance. Developed as well as developing nations share this problem when it comes to farming. Ecologist and wildlife specialist Allan Savory observed:

The United States enjoys the greatest concentration of scientists and wealth ever known in one nation—but she exports more eroding soil annually than all her other exports combined. Wealth, ultimately, means soil. And yet ever-larger farms are said to be "economic" when this is simply not true. The United States claims to be feeding the world when the true position is that the U. S. farmers are bleeding the world with their topsoil losses.

TRINITY COLLEGE LECTURE, DECEMBER 17, 2009

Chemicals in the System

There is further ecological fallout from farming that comes from the chemicals used to substitute for the earth's mechanisms. More than 90% of sprayed pesticides and herbicides go beyond the target area. That means these poisons are in the air surrounding the treated lands, in the waters, and on wildlife. The U.S. Fish and Wildlife Service estimates that 72 million birds are killed each year from pesticides, as well as countless fish. Europe is seeing the decline of at least ten bird species due to pesticides. Frogs and other amphibians are dying off at shocking rates, in part due to pesticides and mutations caused by chemicals.

The effects on humans and other mammals are also being documented. Certain pesticides can alter hormones and disrupt endocrine systems. Pesticide exposure is known to cause birth defects, miscarriages, tumors, childhood leukemias, and cancer, among other health problems. The chemicals tend to accumulate in the body, and sadly, farm workers and children are hit hardest.

Pope John Paul II particularly warned of agricultural toxicity and held industries and individuals responsible, challenging them for reasons of faith and ethics to retool their thinking and detach themselves from a dependency on these environmentally lethal agents:

A more careful control of possible consequences on the natural environment is required in the wake of industrialization, especially in regard to toxic residue, and in those areas marked by an excessive use of chemicals in agriculture.

ADDRESS TO THE XXV SESSION OF THE CONFERENCE OF THE UN
FOOD AND AGRICULTURAL ORGANIZATION, NOVEMBER 16, 1989

Farm corporations and the chemical companies that produce these agricultural supplements argue that these practices *are the only ways* to address the ever-exploding food needs of a growing world population. But what kind of food do they produce? And does this chemically induced food really reach the hungriest people in the world?

Is the Global System Feeding the World?

Built into these industrial food distribution systems are massive productions of food for livestock and processed "junk food" rather than food that nourishes human bodies. The engineered foodlike substances—high in salt, fat, and sugar—are linked to obesity, heart disease, and diabetes in those who consume them, and they drain food budgets while offering little nutrition in return. Emergency room physician J. Matthew Sleethe states that nearly 4 million Americans weigh more than three hundred pounds and 400,000 weigh over four hundred pounds. That's just in one developed country.

There is also a great deal of waste in our food processing and distribution systems, and in what is thrown out uneaten. According to the United Nations Environment Program (UNEP) about one third of the world's food supply (or 1.3 billion tons) is wasted or lost each year, and people in rich countries waste almost as much as is produced in all of sub-Saharan Africa. This worldwide industrial chemical system is a flagrant example of the culture of waste, taking the fertility of the soil and squandering it. Pope Francis emphasized this in the clearest terms possible on World Environment Day 2013: "Throwing away food is like stealing from the table of those who are poor and hungry." He helped launch the UN's anti-food waste campaign to help feed the world.

According to the UN Food and Agriculture Organization estimates, one out of eight people in the world, or 852 million, are suffering from chronic malnourishment, the majority in developing countries. Yet even in developed countries, 16 million have bellies that groan from hunger or bodies that grow poorly for lack of vitamins and nutrition, and this number is on the rise. Though hunger has dropped in some developing countries in Asia, Latin America, and the Caribbean, in Africa hunger has only grown worse as wars and desertification multiply the problems. There, nearly one in four people long for food and the security that they'll have food a week, or two weeks, or a year from now.

Pope John Paul II reprimanded the corporate and market system of agriculture:

> In our own day, farmers collaborating with their Creator can produce enough food for everyone on earth. The fact that the food already available is still not reaching the starving millions is one of the greatest scandals of our age. Such a grave imbalance calls for serious adjustments in the international economic order and greater worldwide cooperation in the production and distribution of food.
>
> ADDRESS TO THE REPRESENTATIVES OF RURAL AUSTRALIA
> AT THE FESTIVAL CENTRE, MELBOURNE, NOVEMBER 30, 1986

Francis picked up John Paul's refrain about the evil of the "culture of waste" with its systems of waste, speculation, and market manipulation built into the global food markets and industrial agriculture. At the 2013 UN Food and Agriculture Conference, he lashed out at those who speculate with food prices, pretending hunger doesn't exist and that they are not responsible:

> It is a well-known fact that current levels of production are sufficient, yet millions of people are still suffering and dying of starvation...
> This is truly scandalous. A way has to be found to enable everyone to benefit from the fruits of the earth, and not simply to close the gap between the affluent and those who must be satisfied with the crumbs falling from the table.
>
> ADDRESS TO THE XXXVIII SESSION OF THE CONFERENCE OF THE
> UN FOOD AND AGRICULTURAL ORGANIZATION, JUNE 25, 2013

The "efficient" worldwide corporate food system—run by the markets, chemicals, and food aid—is efficient for all the wrong goals.

In addition, this farming scenario puts the squeeze on the independent farmers and family farms. With low margins, they often struggle with enormous investments in farm machinery, herbicides, pesticides, and hybrid seeds. Debt is often the result, and when food prices swing up and down, so do bankruptcies. Pope John Paul II worried about the farmers and their losses: "I know too that recently thousands of *American farmers* have been introduced to poverty and indebtedness. Many have lost their homes and their way of life" (HOMILY, MASS FOR RURAL WORKERS IN LAGUNA SECA, MONTEREY PENINSULA, CALIFORNIA, SEPTEMBER 17, 1987).

Similar scenarios play out in other countries. And once the soil has been denuded through the use of herbicides and pesticides, it is difficult to transition out of conventional farming and back to nonchemical practices. It takes five years for the soil to be "clean" again and able to be certified organic. But as the Frantzens have shown, it can be done, and it is worth it.

The present systems of conventional food production, processing, and distribution clearly aren't doing anything close to the job of feeding the world, and even more clearly, they are harming the very systems upon which we, and the rest of life, depend. Without healthy soil and its fertile ecology, the system will collapse. Pope John Paul II asked us to see new possibilities:

> At the beginning of the new millennium, it is an intolerable scandal that so many people are still reduced to hunger and live in conditions unworthy of man. *We can no longer limit ourselves to academic reflections:* we must rid humanity of this disgrace through appropriate political and economic decisions with a global scope.
>
> JUBILEE OF THE AGRICULTURAL WORLD ADDRESS, VATICAN CITY, NOVEMBER 11, 2000

So the systems need to be revamped along moral lines that welcome all to the table and don't poison the food and garden.

Tinkering with the Order of Nature

Pope John Paul II welcomed any agricultural or technological innovation that could help the poor and alleviate suffering, but he pondered the effects of many dangerous practices, noting the warnings the earth itself is giving us about our recklessness:

> The culture of the farming world has always been *marked by a sense of impending risk to the harvest,* due to unforeseeable climatic misfortunes.

However, in addition to the traditional burdens, there are often others *due to human carelessness.*

Agricultural activity in our era has had to reckon with the consequences of industrialization and the sometimes disorderly development of urban areas, with the phenomenon of air pollution and ecological disruption, with the dumping of toxic waste and deforestation. Christians, while always trusting in the help of Providence, must make responsible efforts to ensure that the value of the earth is respected and promoted....

If the world of the most refined technology is not reconciled with the simple language of nature in a healthy balance, human life will face ever-greater risks, of which we are already seeing the first disturbing signs.

HOMILY, MASS FOR THE JUBILEE OF THE AGRICULTURAL WORLD ADDRESS, VATICAN CITY, NOVEMBER 12, 2000

Biotechnology with its genetically modified organisms (GMOs) add another layer of moral dilemma. Inserting genetic matter of one plant species into another or of some virus or animal into a commercial plant greatly concerned John Paul II—especially the mixing of animal and plant DNA. He felt that God had instilled in nature an order that could not be tinkered with without serious repercussions. Profits could never be a sufficient reason for engineering new, unnatural species. Human cloning was nonnegotiable. But even tampering with the building blocks of plant and other animal species was fraught with moral and physical dangers. So he advised the precautionary principle. He warned:

> [B]iotechnologies...cannot be evaluated solely on the basis of immediate economic interests. They must be submitted beforehand to rigorous scientific and ethical examination, to prevent them from becoming disastrous for human health and the future of the Earth."

ADDRESS TO JUBILEE OF THE AGRICULTURAL WORLD, VATICAN CITY, NOVEMBER 11, 2000

Without rigorous scientific and moral testing, caution, and deep humility, extreme damage will occur—as when exotic species are inserted into habitats with unforeseen and disastrous consequences. He was wary of big promises by big corporations.

> Above all the criteria of justice and solidarity must be taken into account. One must avoid falling into the error of believing that only the spreading of the benefits connected with the new techniques of biotechnology can solve the urgent problems of poverty and under-development that still afflict so many countries on the planet.
>
> ADDRESS TO PONTIFICAL ACADEMY OF SCIENCES, 1981

John Paul II taught that "structures of sin"—systems (agricultural, social, economic) stripped of morality and driven by profit, "economic growth," and shareholder value—undermine the planet's quality of life at every level. So for the new millennium, he urged us to transform our industries and systems: "Work in such a way that you resist the temptations of a productivity and profit that are detrimental to the respect for nature" (ADDRESS TO JUBILEE OF THE AGRICULTURAL WORLD, VATICAN CITY, NOVEMBER 11, 2000).

Not Just People and Wildlife Suffering, But Also Livestock

Last, but not least, one must consider the structures of suffering built into the industrial factory farming of animals for efficiency and cheap meat. How can it be moral?

The fact is that inexpensive meat actually encourages many people in developed countries to eat far more meat than is healthy for them, so again, the system encourages gluttony for the few and hunger for the many, while harming the animals involved. What would St. Francis say?

Dr. Sleeth, in the "Food" chapter of his excellent book *The Gospel According to the Earth*, refers back to the in-depth and compassionate mandates for care of livestock in the Bible's first five books. For instance:

> Deuteronomy 25:4, "*You shall not muzzle an ox while it is treading out the grain.*" God states it would be cruel to have one of his creatures helping to make the grain and deny the beast a taste of it. The practice of letting the working ox eat the grain it was grinding represented a

significant economic toll for the farmer. Nonetheless, godly agriculture is not about short-term profits and pleasing shareholders. It is about pleasing the Creator.

Entire books have been devoted to a full discussion of the agricultural and dietary laws under which Israel lived. What it boils down to is that God wants us to treat all his creation lovingly, as he does.

Sleeth opens the chapter by describing the biblical story of Daniel enslaved by the Babylonians, and how he chose to eat a diet of vegetables rather than succumb to the rich diet of his captors, which did not honor God's compassionate precepts about caring for the animals and the land. Daniel did this, as Sleeth observes, because he realized that he could be led astray from God *as much by luxury as by force.* Sleeth directly addresses the issue of factory farming of animals as well:

One biblical concept that flies in the face of modern factory farming is that of not insulting the animal one is eating. For instance, under Jewish law, meat is not served in the milk of its mother (Exodus 23:19). By contrast, factory farming regularly employs the practice of feeding cows, pigs, and chickens the ground up remains of their parents and other animals, irrespective of whether or not God designed the animal as a herbivore. In factory farms, an animal may be kept in a cage where it will never to be able to see the light of day, turn around, lie down, walk or fly. As is seen in the Levitical laws, the use of animals for food is allowed but they must be treated with respect. We are all God's creatures.

As can be expected, the factory farming of chickens, beef, pork, and other animals is not healthy for the workers and neighbors either. With so many animals in such close quarters, enormous amounts of manure are produced, and the air is toxic from the concentrations of it. Animals die in great numbers, so piles of carcasses mount providing another part of the culture of waste. The documentary *Food, Inc.*, created in a lively, personal style, shows the varied perspectives and options in contemporary American agriculture and distribution. The scenes about animal farming, though, are difficult to watch. The stories the farmers and workers tell make clear why we need to rethink our agricultural systems and consider revising them to fit the values of good food and good earth, good service and food for all. For as American corporate farming goes in this global marketplace, so is going the rest of the world.

Pope John Paul II urged us all to aim higher than profit and ease, to aim for what is right in how we farm and *in what we choose to eat and from whom we purchase our food*: "I am therefore happy to encourage and bless those who work to assure that, in the Franciscan spirit, animals, plants, and minerals be considered and treated as 'brothers and sisters'" (POST-ANGELUS: GREETING TO PARTICIPANTS IN TERRA MATER [MOTHER EARTH] SEMINAR, GUBBIO, ITALY, OCTOBER 3, 1982).

Following St. Francis is, of course, a difficult act.

Farming *with* the Order of Nature

How can we live out this Franciscan spirit in today's farming? Consider the Frantzen farm. Organic and sustainable farming seeks to treat land and animals humanely, learning about and working with nature from the soil up. The more holistic and diverse farms are, the more stable, fertile, and economically viable they are for the long term—and the more they encourage the wild ones of God's creatures to inhabit the land and water too. The foods that come from these types of farms tend to go more directly to people who want to eat them, through less processed and healthier food systems, thus offering greater nutrition and fewer junk calories. The soil itself is *lively* as the community of microbes, worms, insects, and minerals are honored and nurtured, so they in turn can nurture the growing of the crops. Valued, the soil is protected with year-round coverage.

There is, of course, an industrial-style organic farming *that takes a huge step in the right direction by putting fewer chemicals into the environment*. However, these large operations tend to be far less diversified and the owners are often separate from the lands they manage. It's not the same as living at the Frantzens' farm as the harvest seasons roll through. "We appreciate home cooking and we like knowing where our food comes from," Irene explains (WWW.ORGANICVALLEY.COOP PROFILE). And of course, they think their natural food tastes better too!

Organic, holistic, and sustainable farmers like the Frantzens know well the truth of what the pope said in Iowa because they witness and help create it daily:

> Is not a miracle worked each day when a seed becomes an ear of corn and so many grains from it ripen to be ground and made into bread? Is not the cluster of grapes that hangs on the branch of the vine one of nature's miracles? All this already mysteriously bears the mark of Christ, since "all things were made through him, and without him was not anything made that was made" (Jn. 1:3).

HOMILY, LIVING HISTORY FARMS, DES MOINES, IOWA, OCTOBER 4, 1979

Costa Rica, in its pledge to go carbon neutral by 2021, is showing that a variety of different kinds of farms can get a nation off the chemical treadmill and feed its people well. The government is training farmers in organic, sustainable, diverse agriculture and having them train others. Environmental reporter Sam Eaton visited a one quarter-acre farmer, Maria Luisa Jimenez, for a typical lunch and sat down to a delicious fresh meal of "pineapple juice, a salad of cabbage, tomatoes and lettuce, sweet yams with syrup, yucca root, rice, tortillas, beans." In the rainy Atlanta coastal area, she says she harvests each month "about 2,000 heads of lettuce, hundreds of cabbages, celery, cauliflower, cucumber—the list goes on. And it's all organic." Or as she says, *organoponic*. She relies on a fertilizer mix of "dried rice husks, coconut fiber, composted cow manure and biochar, a charcoal-like material made from agricultural waste that would otherwise be burned" (WWW.THEWORLD.ORG WEB SITE, "CARBON-NEUTRAL LUNCH," JULY 1, 2013).

More than 550 Costa Rican farm families have learned and begun to employ these techniques, and as Jimenez boasts, "'You'd see that everyone is in the same position as we are. They have a lot of food, they have many ways to feed themselves in a healthy manner while caring for the environment.'"

Eaton also stopped in on a hog farm that uses a big balloon-like methane biodigester over the manure pit, which produces energy for running other parts of the farm. The farmer also plants trees for fence posts and applies an accounting template to keep track of carbon use and emissions.

Then there's the industrial Dole banana plantation. Because of the government's new carbon-neutral goals, everything from agriculture to energy and industry—even the multinational, industrial-style Dole Food Company—has to aim for the big carbon zero too. So Dole has reduced its fertilizer use to BB-sized, time-release pellets; it recycles the plastic inserts used to protect the bananas; and it has cut water use at its processing plant by more than three fourths. To cover the carbon emitted in shipping, they've set aside three times the size of the plantation in native forest.

As Sam Eaton notes, the various "set-asides" will cause a little tricky accounting to effectively tabulate the net carbon emission, but it's a start. If all nations could follow this lead in agricultural and industrial policies, the definition of farming could return more swiftly and easily to that of a vocation of service, in line with the Garden of Eden first farm family.

Instead of developed nations exporting chemical and genetically modified seed to developing countries, perhaps the developing countries like Costa Rica who are experimenting with transforming all levels of their systems can teach a thing or two to the developed nations who are content with the status quo.

The Struggles of Subsistence Farming

Of course, much of the world's farming is far removed from organic and sustainable farming networks, or large tractors, combines, and herbicide applicators. Pope John Paul II witnessed the struggles of subsistence farmers around the world, especially in the Sahel. Their challenges were not related to misusing abundance but to handling scarcity. In his Jubilee of the Agricultural World address, the pope called attention to their plight and to the responsibility of all to join in solidarity to look for better more sustainable methods for subsistence farmers that build up the soils, animals, and people rather than deplete them. And to end war. Armed conflicts and land mines make farming all the more difficult. People have to eat, so they can't just decide not to tend their fields because a war is going on:

> Entire peoples, who depend primarily on farming in economically less developed regions, live in conditions of poverty. Vast regions have been

devastated by frequent natural disasters. And sometimes these misfortunes are accompanied by the consequences of war, which not only claims victims, but sows destruction, depopulates fertile lands and even leaves them overrun with weapons and harmful substances.

..... We must contribute to a culture of solidarity which, at the political and economic level, both national and international, encourages generous and effective initiatives for the benefit of less fortunate peoples.

JUBILEE OF THE AGRICULTURAL WORLD ADDRESS, VATICAN CITY, NOVEMBER 11, 2000

Besides having to farm alongside land mines and face the risks of kidnapping and conflict, many people endure invading deserts and depressed crop prices and demand due to global products hitting their home markets. To try to compete, some turn to Western-style chemically based farming and patented hybrid or GMO seeds, which can increase their gains temporarily but bring with them the additional costs of pesticides, fertilizers, and annual seeds, leading to the same financial and ecological problems as in developed countries. In contrast, the UN Food and Agricultural Organization has found that mixed sustainable farming using crops and livestock on the same farm is frequently the most productive form of agriculture—for example, diversified home compounds in Nigeria yield five to ten time as much per hectare as monocrop fields.

So what should subsistence farmers do, especially where their lands are turning to desert and drought has stopped growth?

Native Grasslands, Livestock Herds, and Holistic Management

Allan Savory, originally from Zimbabwe, has watched farmers in his homeland and various regions of Africa struggle over the years with the loss of their grazing lands and soil fertility. Studying the history and natural rhythms of the land, Savory developed a holistic way to fight desertification through native grassland farming. He encouraged villagers to gather their livestock into large communal herds and move them through sequential set-asides of land, like nomadic herders have done for ages. This mimics having predators move wild herds of gazelles, zebras, buffalo, and other grazers. Imitating nature through holistic management can bring back the grasslands from desert through the "poop and stomp method."

In a 2013 TED Talk, "How to Green the Desert and Reverse Climate Change," Allan Savory received a standing ovation for his plan to help call back the five billion hectares of grasslands and savannahs deserts, while setting free people who are now dependent on international food aid to feed themselves. Based on present-day carbon estimates, this enormous addition of native grass matter spread across the planet would sequester carbon to the point of stabilizing the climate (especially in coordination with equally ambitious reforestation programs, such as the Green Belt Movement), while capturing moisture for rivers, wells, gardens, and wildlife.

He showed before-and-after photographs of this holistic grazing and animal care, where wildlife has multiplied—proof that all life benefits from the return of these diversified native grasslands that can beat the predatory desert. His successful Operation Hope project in Zimbabwe, a Buckminster Fuller Challenge Finalist, re-grassed 6,500 acres of denuded lands, reversing desertification while multiplying livestock by 400% and wildlife by substantial proportions. These agricultural methods also increased the community's self-sufficiency, standard of living, and cohesiveness, because all the villagers had to work together to consolidate livestock into communal herds and keep them moving over common lands—"embracing local traditions and culture, and opening opportunities for younger generations to stay on the land, and make a meaningful living while leading prosperous lives," as stated in the Savory Institute's 2012 Annual Report. In summation, "This is the difference between a world in harmony, and a world in chaos."

Or, as John Paul II put it, "peace with all of creation."

Restoration Agriculture

So it's not enough just to stop degrading the earth, we need to help heal and restore it as well. Holistic agriculture and restoring grasslands with livestock is a form of the growing branch of agriculture that fits Pope John Paul II's ethical tenets of restoration. The permaculture movement uses perennial trees and plants, like native grasses, flowers, berries, and vines to keep the soil covered, and draws its harvests from them. Its principles entail care of the earth, care of the people, and sharing of the surplus.

At New Forest Farm near Viola, Wisconsin, Mark and Jen Shepard restored lands destroyed by industrial farming by planting chestnut, hazelnut, walnut, apple, cherry, pear, and other trees in swirling designs that fit the ridged contours of the land. In between the rows, they grow perennial asparagus as well as winter squash. Their hogs, highland cattle, and chickens rotate through the rows in management sequences, eating as they go. Keyline irrigation swales and pocket pools for catching rain hold and direct the water while offering frogs, ducks, and other wildlife hospitable places to reside. A wind turbine and solar panels on the roofs supply off-the-grid energy needs. Wild grapes and berries, along with all kinds of wildlife call the Shepards' farm home. It has become a showplace for "Restoration Agriculture"—demonstrating how to call forth abundance from the land naturally while restoring it at the same time.

"There are two problems with agriculture—even organic agriculture," Mark Shepard told *Grist* reporter Jake Olzen wryly. "You are either trying to keep something alive that wants to die, or you are trying to kill something that wants to stay alive." In contrast, Shepard prefers farming that doesn't require annual plantings: agroforestry (farming nut and fruit trees) and perennials (such as asparagus and grapes, that return yearly on their own), so he can harvest the fruits of the nature's labors, not his.

This is another way of living out Pope John Paul II's definition of conservation—becoming part of the rhythms of creation as a partner and co-creator with God, nurturing the natural systems of abundance that are continually renewing themselves:

> Having created the cosmos, God continues to create it, by maintaining it in existence. Conservation is a continuous creation....
>
> Divine Providence is expressed especially in this "conservation," namely, in maintaining in existence all that has had being from nothing. In this sense Providence is a constant and unending confirmation of the work of creation in all its richness and variety. It implies the constant and uninterrupted presence of God as Creator in the whole

of creation. It is a presence that continually creates and reaches the deepest roots of everything that exists.

GENERAL AUDIENCE OF MAY 7, 1986

The Underappreciated Profession

Farming is a significant vocation to work so directly with God, and see the visible effects almost daily. Unfortunately, it is not valued that way in most developed societies.

As farming has become more corporate, with fewer and fewer family farms in operation, the profession has become more isolated from those who eat the food. Pope John Paul II witnessed the resulting lack of understanding of farmers' challenges and worth. He reminded the world how much we depend upon farmers:

> When this sector is underappreciated or mistreated, the consequences for life, health and ecological balance are always serious and usually difficult to remedy, at least in the short term.
>
> ADDRESS TO JUBILEE OF THE AGRICULTURAL WORLD, VATICAN CITY, NOVEMBER 11, 2000

He called upon the world to look upon farmers with respect and to give thanks to God and to them for their indispensable work:

> It is beautiful and proper to thank God for the gifts received in the course of the year and to be grateful to the men and women who reap them from the earth with their work. Farmers, not often recognized in industrial societies, merit instead universal thanks for the essential service they render to the whole human family.
>
> ANGELUS, NOVEMBER 10, 2002

Part of giving thanks is to purchase their foods more directly, so more of the money stays with them and you can tell them face-to-face that you like their harvests. There are farmers' markets and local producers of organic or sustainable eggs, meats, fruits, vegetables, and nuts in more and more places. In Community-Supported Agricultural (CSA) shares, you can purchase part of the varied crop as it comes in and even help in the farming. U-pick operations, especially for fruit and berries, give you a chance to taste sun-ripened produce. The food may be a little more expensive, but it is also healthier and contributes to the service of God's kingdom, keeping this planet healthy, fertile, and beloved. Community bonds form between rural and urban dwellers. Such purchasing acts help us all to be more thoughtful, judicious eaters and budgeters, which is not only beneficial to the planet but also helps to keep obesity at bay.

And if we grow some of our own herbs and vegetables, and maybe even raise a few chickens for eggs, we get a little taste of what is involved, watching things grow and tasting the freshest food possible (and also the trials and disappointments of failed crops!). Pope John Paul II blessed us all in this effort, offering his special appreciation and blessing to those who tend the earth so the rest of us may live:

> Only those who till the land can really testify that barren earth does not produce fruit, but when cared for lovingly, it is a generous provider. With gratitude and respect, I bow before those who for centuries made this land fruitful by the sweat of their brow. . . .
>
> With the same gratitude and respect, I also speak to all who are engaged today in the hard work of tilling the land. May God bless the work of your hands!
>
> LITURGY OF THE WORD, ZAMOŚĆ, POLAND, JUNE 12, 1999

*Let it not be said that the fair and equitable recognition of
Aboriginal rights to land is discrimination.
To call for the acknowledgment of the land rights of people who
have never surrendered those rights is not discrimination.
Certainly, what has been done cannot be undone.
But what can now be done to remedy the deeds of yesterday
must not be put off till tomorrow.*

ADDRESS TO THE ABORIGINES AND TORRES STRAIT ISLANDERS,
ALICE SPRINGS, AUSTRALIA, NOVEMBER 29, 1986

9. Indigenous Peoples, Lands, and Environmental Racism

On every inhabited continent, indigenous peoples are struggling to maintain and protect their language, culture and rights from the encroaching outside world. Many of their lands are being logged and mined, their wildlife poached, and genetic material stolen from their lands despite their protests. Roads and pipelines are built through their territories. Hazardous and polluting materials are being dumped in their "backyards." They are commonly raped, murdered, or beaten with little criminal justice for their perpetrators, who are generally outsiders. In all key ecological issues of today, indigenous people are deeply affected.

Despite the oppression and discrimination, these indigenous communities struggle to maintain a closer, kinder relationship to the earth, their poor, and their elders. And many are using creative and dynamic ways to try to protect their lands and build their communities back from histories of genocide, imprisonment, invasions, land and natural resource robbery, consumerism, and other assaults from the outside—while still engaging in contemporary education, technology, and civilization.

In each continent and nation he visited, Pope John Paul II made it a point to meet with its indigenous people. In admiration, Pope John Paul II told the First Nations peoples of Canada and the United States:

For untold generations, you, the Native peoples, have lived in a relationship of trust with the Creator, seeing the beauty and the richness of the land as coming from His bountiful hand nd as deserving wise use and conservation. Today you are working to preserve your traditions and consolidate your rights…

HOMILY AT MASS FOR THE NATIVE PEOPLES
OF CANADA, FORT SIMPSON, SEPTEMBER 20, 1987

He acknowledged their courage against the structures of racism, imperialism, and greed assaulting them, and honored them for the models and values they offered in contrast to the wider society's materialism. In 1986, in a Mass for the aboriginal peoples of Australia, he observed in his homily:

For thousands of years, you have lived in this land and fashioned a culture that endures to this day. And during all this time, the Spirit of God has been with you. Your "Dreaming"—which influences your lives so strongly that no matter what happens you remain forever people of your culture—is your way of touching the mystery of God's Spirit in you and in creation. You must keep your striving for God and hold on to it in your lives. …

Your culture, which shows the lasting genius and dignity of your race, must not be allowed to disappear. *Do not think that your gifts are worth so little that you should no longer bother to maintain them.* Share them with each other and teach them to your children. Your songs, your stories, your paintings, your dances, your languages, *must never be lost.* …

For thousands of years, this culture of yours was free to grow without interference by people from other places. You lived your lives in spiritual closeness to the land, with its animals, birds, fishes, waterholes, rivers, hills, and mountains. Through your closeness to the land, you touched the sacredness of man's relationship with God, for the land was the proof of a power in life greater than yourselves. …

ADDRESS TO THE ABORIGINES AND TORRES STRAIT ISLANDERS,
ALICE SPRINGS, AUSTRALIA, NOVEMBER 29, 1986

Though indigenous peoples around the world are as diverse as the world's different nations and regions, they share in common an inherent relationship with nature and the lands they live with. Theirs is not a sentimental tie, but an essential one of identity, and they are tied not just to the land, but to all the life on it, as embodied in the Dakota Sioux saying *Mitakuye Owasin*—"We are all related" or "All my relations." This is a quality and way of life John Paul admired deeply and asked all people to imitate. He told the aboriginal Australians people in 1986 that one of the reasons he held them in such high honor was because, "You did not spoil the land, use it up, exhaust it, and then walk away from it. You realized that your land was related to the source of life."

The World Needs to Follow Their Model

Early in his papacy, in a radio address to the First Nations peoples in Canada, he told them that because of their close kinship with all living things, a kind of special responsibility and vocation had been laid upon them to lead the world by example:

> You are called to responsible stewardship and to be *a dynamic example* of the proper use of nature, especially at a time when pollution and environmental damage threaten the earth.
>
> RADIO AND TV MESSAGE TO NATIVE PEOPLES OF CANADA,
> YELLOW KNIFE, SEPTEMBER 18, 1984

The pope asked the world to try to see Creation, or nature, with the eyes of the indigenous people who view themselves as intimately, reverently, and responsibly related to it.

> A central element of indigenous culture is their attachment and nearness to mother earth. You love the land and want to keep in contact with nature.
>
> I join my voice to yours and others asking for the adoption of strategies and means to protect and preserve nature. Due respect for

the environment must always be held above purely economic interests or abusive exploitation of the resources of land and sea.

MESSAGE TO THE INDIGENOUS PEOPLES OF SANTO DOMINGO, OCTOBER 12, 1992

The pope was equally inspired by their human ecology—how they cared for each other and their appreciation for those things that cannot be bought and sold. The indigenous communities he encountered cared for the poor, orphaned, and elders among them; valued family and their kinship ties over possessions; and carried with them important life wisdom and skills, as well as concrete biological, intellectual, creative, and spiritual knowledge. He addressed the Christian indigenous people of the northern regions of North America as well as all native people:

> Over the centuries, dear American Indian and Inuit peoples, you have gradually discovered through your cultures' special ways of living your relationship with God and with the world, while remaining loyal to Jesus and to the Gospel.
>
> Continue to develop these moral and spiritual values: an acute sense of the presence of God, love of your family, respect for the aged, solidarity with your people, sharing, hospitality, respect for nature, the importance given to silence and prayer, faith in providence.
>
> *Guard this wisdom preciously.*

ADDRESS TO AMERINDIANS AND INUITS, SAINTE-ANNE DE BEAUPRÉ, CANADA, SEPTEMBER 10, 1984

Dr. Charles Eastman, or Ohiyesa, was a Dakota Sioux who earned a medical degree from Boston University Medical School in 1890. When he went to share his newfound faith in Christianity with the Dakota and Lakota, many of the people looked at him in puzzlement. From his teachings, Jesus sounded to them like an Indian. John Paul, too, saw that the values of the indigenous peoples are the same as Jesus's spiritual values, whether or not the people were Christian:

> Through his Gospel, Christ confirms the native peoples in their belief in God—their awareness of his presence, their ability to discover him

in creation, their dependence on him, their desire to worship him, their sense of gratitude for the land, their responsible stewardship of the earth, their reverence for all his great works, their respect for their elders.

The world needs to see these values...made incarnate in a whole people.

LITURGY OF THE WORD WITH THE NATIVE PEOPLES OF CANADA, HURONIA, CANADA, SEPTEMBER 15, 1984

For many in contemporary Christian culture, this concept of following the model of indigenous peoples seems contrary to the traditional self-concept that Christians came to "civilize" native cultures. However, such a view represents a shallow understanding of the richness of the native beliefs and cultures and of the history of Christianity. Jesus was a Jew who practiced the earth-care rituals and understandings of the indigenous Jewish beliefs. As Christianity moved throughout the world, it often integrated the earth-care and spiritual connectedness of the peoples who embraced the Christian conversion, such as the Celts in Ireland, and the varied native peoples in Mexico, Latin America, and North America. (Sadly, too often for the native peoples, the integration was forced upon them with brutality.) Christian cultures that have integrated the full wisdom of both spiritual paths are among the most spiritually vibrant in the world, dedicated to Christ as well as caring for families and for their place in the earth's garden.

With the world in ecological crisis, John Paul II asked all the people of the globe to learn from these native cultures, to recapture the neglected ecological wisdom, and like them, honor God's spirit force revealing itself everywhere—in the dawn's glistening and moon's iridescence, a flower pushing up through cracked pavement, a mom talking patiently to a crying toddler, an old woman kissing her equally old husband on the cheek with a dimpled smile of appreciation.

Cure for Romanticization: Having a Voice

The danger, of course, in Pope John Paul II's counsel and observations is that they could lead to a monolithic, stereotyped, romantic version of the varied indigenous peoples, like the old concept of the "noble savage." This stereotype has undermined the essential diversity of indigenous people, their human failings and humor, their individualized capabilities and unique perspectives. John Paul II was very careful not to do this and to study something of the history and culture of each community and religious culture before he arrived. He also fought against the pressure to "keep them as they are" so they can fulfill glamorized spiritual and environmental roles assigned to them by outsiders. The pope feared that corporate, religious or nonprofit organizations or governmental agencies would try to take over this role, prioritizing their view of a region's ecological and developmental needs over the right of local indigenous people to determine their own future and speak for themselves: "If the protection of the environment were to be made an end in itself, there is the risk that new, modern forms of colonialism will arise that would injure the traditional rights of communities...." (MESSAGE FOR THE 23RD WORLD DAY OF TOURISM, JUNE 24, 2002).

The Indigenous Environmental Network (IEW) has this as its first principle of Environmental Ethics—that no one should ever speak *on behalf* of indigenous people; *they should speak for themselves*. Unfortunately, indigenous peoples and their voices continue to be marginalized legally and politically in most of the countries in which they live. Whether in Africa, Asia, Australia, the Americas, or the islands in the Pacific, the voices of indigenous people are rarely attended to with the same respect or attention as those of government agencies, corporations, development companies, or economic interests. For instance, the Enbridge Oil Pipeline and proposed Keystone Pipeline both traverse the lands of First Nations in Canada and American Indian nations in the United States, harming their water supplies, soil, and wildlife. Yet the media simply ignores them. Courts often dismiss or rule against indigenous peoples in nations around the globe.

The pope scolded the world for these injustices:

> Every attempt to marginalize the indigenous peoples must be eliminated. This means, first of all, respecting their territories and the pacts made with them; likewise, efforts must be made to satisfy their legitimate social, health, and cultural requirements. And how can we overlook *the need for reconciliation between the indigenous peoples and the societies in which they are living?*
>
> POST-SYNODAL APOSTOLIC EXHORTATION, *ECCLESIA IN AMERICA*, 1999

Popes Benedict XVI and Francis followed John Paul II's lead on meeting with indigenous peoples of the world and speaking out on their behalf. (Benedict, though, had a mixed response, as some of his words, which were later modified, were taken as offensive.) In 2013 in Brazil, Pope Francis met with various members of Brazil's native communities. He joined his voice to their fight to save Amazonia and defend their rights to their lands and water. When they tried to kneel and kiss his ring, he pulled them up, blessed their foreheads, and embraced them. A young man presented him with a feathered headdress, which he wore with a delighted smile, reminding the world of the many times John Paul II beamed as he put on a traditional headdress offered as a gift.

Striving to Start a New Relationship Through Forgiveness and Respect

Even as Pope John Paul II rebuked the world for racism and oppression against indigenous peoples, he was acutely aware of the unsavory role of the Catholic Church and other Christian denominations in the histories of indigenous peoples throughout the world. None should *ever have used violence, force, shame, or oppression* in the name of God! In contrast were the humble missionaries, many of whom were amazingly courageous and self sacrificing, learning the peoples' ways and languages, and offering their knowledge of Christ and medicine as gifts, *unfettered gifts*, gifts to be used *as the people desired, not as acts of domination or disrespect.*

In his tour of the United States in August 1987, Pope John Paul II attended the National Tekakwitha Conference in Phoenix, which was named after the first Native American to move to sainthood in the Catholic Church—St. Kateri Tekakwitha, from the *Kanien'kehake*, the People of the Flint, also known as Mohawks. (In 1980, the pope had beatified Kateri, and in 2012, Pope Benedict continued the process and proclaimed her a saint—a patron of ecology and the environment, to join St. Francis of Assisi.) At this conference of Native American Catholics, John Paul II addressed the people gathered, thanking them for their invitation:

> I have greatly looked forward to visiting with you, the original
> people of this vast country.... As your representative spoke, I traced

in my heart the history of your tribes and nations.... Here your for-
bears worshipped the Creator and thanked him for his gifts. In contact with the forces of nature, they learned the value of prayer, of
silence and fasting, of patience and courage in the face of pain and
disappointment...

*The early encounter between your traditional cultures and the European
way of life was an event of such significance and change that it profoundly
influences your collective life even today.* That encounter was a harsh and
painful reality for your peoples. The cultural oppression, the injustices,
the disruption of your life and of your traditional societies must be
acknowledged....

Unfortunately, not all members of our Church lived up to their
Christian responsibilities.... Now, *we are called to learn from the mistakes
of the past and we must work together for reconciliation and healing*, as brothers and sisters in Christ....

I encourage you... to work *to preserve and keep alive your cultures,
your languages, the values and customs* which have served you well in the
past and which provide a solid foundation for the future.

Your customs that mark the various stages of life, your love for
the extended family, your respect for the dignity and worth of every
human being, from the unborn to the aged, and your stewardship
and care of the earth: these things benefit not only yourselves but the
entire human family....

PHOENIX, ARIZONA, SEPTEMBER 14, 1987

The pope encouraged the extended American public and the American
Catholic Church to listen more to the Native American Catholics and stop
the racism:

From the very beginning, the Creator bestowed his gifts on each people. It is clear that stereotyping, prejudice, bigotry and racism demean
the human dignity which comes from the hand of the Creator and
which is seen in variety and diversity....

We should all be grateful for the growing unity, presence, voice
and leadership of Catholic Native Americans in the Church today....

> Here too I wish to urge the local churches to be truly "catholic"
> in their outreach to native peoples, and *to show respect and honour for*
> *their culture and all their worthy traditions....*
>
> As Catholic Native Americans, you are called to become *instru-*
> *ments of the healing power of Christ's love*, instruments of his peace.

ADDRESS TO THE NATIONAL TEKAKWITHA CONFERENCE,
PHOENIX, ARIZONA, SEPTEMBER 14, 1987

After his address, a holy man and healer of the Mescalero Dine Apache, Sidney Baca, blessed him in a traditional way, with prayers, sweet grass, and an eagle feather. Baca then presented the eagle feather to the pope, who accepted it gratefully.

Following Words with Actions

Still, as indigenous people know, words are cheap. Apologies can mean much or nothing, depending on the follow-through. Treaties have proven weightless, drifting on the currents of time. Sometimes they bob to the top and people pay brief attention to them; most often they are drowned in the storms of politics and greed. Pope John Paul II knew this well. He had seen people struggling for justice on their lands, struggling to be heard, for treaties to be remembered, for their lands to be their own. He knew that they were not just living in a state of harmonious nirvana with the land, but striving daily just to hold on to what they have and survive, while engaging with the contemporary world.

For instance, currently less than 10% of the White Earth Reservation lands of the Anishinaabe (also known as Ojibwe/Chippewa) of Northern Minnesota are held by tribal members, and more than three fourths of their people live "as paupers and refugees" in the Twin Cities or other areas. Their White Earth Land Recovery Project (WERP), founded by Winona LaDuke, is a nonprofit tribal organization that facilitates "recovery of the original land base of the White Earth Indian Reservation, while preserving and restoring traditional practices of sound land stewardship, language fluency, community development, and strengthening our spiritual and cultural heritage"

(WERP WEBSITE). (Educated in economics at Harvard and Antioch Universities, LaDuke came to the reservation to serve her people.)

On their lands, they are harvesting wild rice, berries for teas and jellies, traditional corn, maple syrup, birchbark, and bison to sell on their Native Harvest website. "We work to continue, revive, and protect our native seeds, heritage crops, naturally grown fruits, animals, wild plants, traditions and knowledge of our indigenous and land-based communities; for the purpose of maintaining and continuing our culture and resisting the global, industrialized food system that can corrupt our health, freedom, and culture through inappropriate food production and genetic engineering." This is precisely the kind of positive healing project of lands, people, and economies that Pope John Paul II believed in and urged all Catholics and other Christians and the world at large to encourage, support, and nurture. He told Native Americans to keep pursing their unique vocation, affirming challenging works like those of Native Harvest and the White Earth Recovery Project and supporting those whose hope may be flagging:

> All consciences must be *challenged*. There are real injustices to be
> addressed and biased attitudes to be changed. But the greatest chal-
> lenge is to you yourselves, as Native Americans. You must continue to
> grow in respect for your own inalienable human dignity, for the gifts
> of Creation and Redemption as they touch your lives and the lives of
> your peoples. You must unyieldingly pursue your spiritual and moral
> goals. You must *trust* in your own future.
>
> ADDRESS TO THE NATIONAL TEKAKWITHA CONFERENCE,
> PHOENIX, ARIZONA, SEPTEMBER 14, 1987

To Save Indigenous People, One Must Protect Their Lands

These people cannot live out their unique vocation, however, if they are parted from their lands. Pope John Paul II criticized international, national, and local bodies for not working enough to protect native rights. He demanded far stricter laws, policies, and enforcement to protect indigenous communities' rights to their own lands, forests and waters, religious freedom, and their ability to speak for themselves to determine their own futures. He explained why they must all work for native land preservation:

> Certain peoples, especially those identified as native or indigenous, have always maintained a special relationship to their land, a relationship connected with the group's very identity as a people having their own tribal, cultural, and religious traditions.
>
> When they are deprived of their land, they lose a vital element of their way of life and actually run the risk of disappearing as a people.

WORLD DAY OF PEACE 1989 MESSAGE,
"TO BUILD PEACE, RESPECT MINORITIES," DECEMBER 8, 1988

Because of so many devastating infringements on the rights and lands of native peoples across the globe, the United Nations spent twenty-four years, with the pope's support, developing a "Declaration on the Rights of Indigenous Peoples." After he died, it was made official in 2007. Thankfully, the four countries at first reluctant to sign on—Canada, Australia, New Zealand, and the United States—have since approved it. This declaration has not solved all problems or inequities, especially in the areas of environmental racism and injustice, but it's a statement of the standard by which all should be judged.

Environmental Racism Against Indigenous Peoples

Too few people in the world really understand the depth of the racism, violence, pollution and destruction perpetrated against poor indigenous people and how serious a problem it is. In 2010, the International Indigenous Women's Symposium gathered in California to address the environmental injustice and violence visited upon them and their lands. They came from North America, Latin America, the Arctic, the Caribbean, and the Pacific, to write their own "Declaration for Health, Life and Defense of Our Lands, Rights, and Future Generations." In it, they spoke out for the rights of their unborn and children, who are being harmed by the toxic pollution of others—and their absolute determination to fight both the pollution and the ecological racism at the root of it, which harms their families, children, economies, lands, brother and sister animals, and the future generations:

> We are traditional healers, midwives, youth and community organizers, environmental and human rights activists, teachers and traditional and cultural leaders. We are daughters, sisters, mothers, aunties, grandmothers and great grandmothers, youth and elders, members of great Nations who have always stood firm to defend our lands, our Peoples and our cultures....
>
> We have come together at this Symposium to share our information about the negative impacts of mining and drilling, mercury contamination, nuclear and uranium testing, processing and storage, pesticides and Persistent Organic Pollutants (POPs), military dumping, toxic waste incineration, desecration of sacred sites and places, introduction of Genetically Modified Organisms (GMO) and foods and harvesting of our genetic materials....
>
> These imposed, deplorable conditions violate the right to health and reproductive justice of Indigenous Peoples, and affect the lives, health and development of our unborn and young children. They seriously threaten our survival as Peoples, cultures and Nations. They also violate our rights as Indigenous Peoples to subsistence, spiritual and cultural survival, self-determination and free, prior and informed consent (FPIC).
>
> As Indigenous Peoples, and as the defenders of our future generations, we have vocalized our opposition to these forms of contamination of our homelands, air and waters for generations in many different regions, but far too often we are ignored. We have also shared our strategies and ideas about how to address these situations in our communities and around the world.

They summed up the problem: "All living beings are being misused and poisoned by corporations, States and their Territories, based on foreign and colonial concepts that disregard the sacredness of life." And they listed the devastating health issues, which are all at their heart pro-life issues:

- contamination of mothers' breast milk at 4 to 12 times the levels found in the mother's body tissue;
- elevated levels of contaminates such as POPs and heavy metals in infant cord blood;
- disproportionate levels of reproductive system cancers of the breasts, ovaries, uterus, prostate, and testicles, including in young people;
- elevated rates of respiratory ailments such as asthma and lung disease;
- high levels of leukemia and other cancers in infants, children, and youth;
- rare, previously unknown forms of cancer among all ages in our communities;
- devastating, and in many cases, fatal birth defects known to be associated with environmental toxins such as nuclear waste, mining, and pesticides, including the increasing birth of "jelly babies" in the most contaminated areas;
- developmental delays, learning disabilities, and neurological effects on babies and young children which have lifelong impacts, associated with prenatal exposure to mercury, pesticides, and other environmental toxins; and
- increasing numbers of miscarriages and stillbirths, and high levels of sterility and infertility in contaminated communities.

They also decried issues related to the mining and the growing of genetically modified foods on their lands and denounced the sexual violence and exploitation of indigenous women. They set out an action plan of education, training, and support for indigenous communities, and policies for corporations, government agencies, and international bodies.

In the United States, many native and nonnative people have bonded together for strength and support in the Honor the Earth environmental organization, which positions itself as "a voice for the earth...a voice for those not heard." "Native people have borne the brunt of America's past energy policy," says the Anishinaabe activist Winona LaDuke, who is also cofounder and director of the program, "from uranium mining in the southwest to massive hydroelectric projects in the subarctic. It is time for energy justice, and it is time for a new energy policy."

Honor the Earth seeks not only to protect their members' own scattered native lands, but they also strive to protect the supportive ecological systems—protesting, for example, nuclear waste storage and pipelines, such as Enbridge and Keystone, across their lands, fossil fuel expansion, and many other threats. They are forging wide-ranging energy solutions on Indian reservations, establishing wind and solar power stations like beads on an energy necklace strung across the nation to sell power to the larger grid and reduce the need for coal and nuclear plants.

In other countries, similar indigenous networks, research, and grant-making efforts are underway. Consider the First Nations Development Institute in Canada or the National Indigenous Organization of Colombia, or the Asian Indigenous Women's Network. The Indigenous Environmental Network (IEN) or the Indigenous Women's Network (IWN), and the Indigenous Women's Biodiversity Network (IWBN), are all organizations that help coordinate, give voice to, and support all their ecological efforts internationally.

The Darker You Are, the More Racism You Face?

In many places on earth, the browner you are, and the closer you live to the earth, the more you are despised. In former colonies, "white supremacy" and favoring gradations of lighter-skinned people (though that is not the only form of racism) is a plague that has devastated the human family, and continues to do so. Consider the urban African American and Latino neighborhoods in the United States, where trash incinerators are commonly built, or the Pacific island communities, where nuclear testing occurs. The World Bank's chief economist for Africa, Shantayanan Devarajan, stated that the cause of much of the inequity and the usurpation of indigenous land rights is racism. This confirms the findings of earlier studies and international conferences. Calling all forms of racism "a serious offence against God," Pope John Paul II stated that *"Every upright conscience cannot but decisively condemn any racism, no matter in what heart or place it is found."* He added:

> To oppose racism, we must practice the culture of reciprocal acceptance, recognizing in every man and woman a brother or sister with

whom we walk in solidarity and peace. There is need for a vast work of education to the values that exalt the dignity of the human person and safeguard his fundamental rights.

ANGELUS, AUGUST 26, 2001

The most dramatic, public expressions of God's rejection of racism are in the apparitions of the Virgin Mother Mary around the world. She appears to poor and devout people, often children and nuns, asking them to pray for peace and the conversion of hearts to God. When she gave her image to the indigenous man Juan Diego on his cloak at Guadalupe, it was not of a white, blond Mary, as in many European renderings, but of a brown-haired, brown woman, dressed in clothing and colors significant to the native local people, with symbols of creation all around her. When she gave the gift of a statue of herself to a praying fisherman in Brazil, it came out of the river and was made of unpainted reddish-brown clay. When she appeared to poor children in the mountain village of Kibeho, Rwanda, prior to the genocide, she said she was "Nynia wa Jambo," meaning "Mother of the Word," a synonym for "Umubyeyi w'Imana," which means "Mother of God." She was black as the girls she visited, and she appeared often with fields of flowers.

Rwanda had a problem of very high population and declining land fertility from unsustainable land practices, deforestation, and soil erosion, leading to desertification. Its history had included Belgian colonialism, which favored the lighter-skinned Tutsi herders over the Hutu farmers. In her apparitions in the 1980s, Mary begged Rwandans to pray for peace and to change their hearts or there would be horrendous carnage, "a river of blood," and she showed the youth visions of what might come. She told one of the girls, Marie Claire Mukangango, on March 27, 1982:

> "When I show myself to someone and talk to them, I want to turn
> to the whole world. If I am now turning to the parish of Kibeho, it
> does not mean I am concerned only for Kibeho or for the diocese of
> Butare or for Rwanda, or for the whole of Africa. The world is bad.
> The world rushes toward its ruin. It's about to fall into an abyss. The
> world is in rebellion against God. Many sins are committed. There is

no love and no peace. If you don't repent and convert your hearts, all will fall into an abyss."

Mary's appearances were pleas and warnings of love. In all her apparitions, her messages have been consistent: that she loves all people, God loves them even, and we need to abandon racism and hatred and violence, turning to prayer and fasting, kindness, and forgiveness to bring about a world at peace with all people and creation.

In his visit to Rwanda in 1990, Pope John Paul II told Rwandans to turn to Christ, and to listen to his mother as "a simple and sure guide" counting on her care, asking for help to overcome racial and political divisions. Then drought, austerity from the World Bank, and a drop in prices for their main exports, coffee and tea, created increased hunger and political unrest.

Sadly, Rwanda as a whole did not embrace Mary's message. In 1994 and 1995, Kibeho became the sight of two massacres, in which some of the visionaries were killed. Marie Claire was killed trying to protect her husband at another location. Another girl who said she had the visions, Immaculee Ilibagiza, was a Tutsi who was by then studying to be an engineer. She survived because a Hutu priest hid her and some other women for ninety-one days. She has written three books on her experiences, and has become an international speaker on faith, forgiveness, and overcoming racism.

The Cultural and Psychological Effects of the Environmental Racism

As has been seen in other areas, violence against the land is associated with violence against people. For instance, in 2004, the U.S. Department of Justice completed a ten-year study entitled "American Indians and Crime." It found a "disturbing picture of victimization of American Indians and Alaska Natives." It assessed that one in ten American Indians or Native Americans have been a victim of violence, twice the percentage of African Americans and two and a half times that of white Americans. It also noted that 70% of the reported violent attacks are perpetrated by non-Indians. Lawrence A. Greenfield, coauthor of the study, noted that "American Indians experience a much greater exposure to violence than other race groups, and it is from non-Indian people against them." A study conducted in 2010 found that Native American and Native Alaskan women are two and a half times more likely to be victims of sexual violence than the general populace, and one out of three are raped in their lifetime. This is not uncommon for indigenous women around the world, and again, much of this violence is perpetrated by outsiders.

In his address to the aboriginal peoples of Australia, John Paul II expressed his sorrow for the violence and racism:

> Many of you have been dispossessed of your traditional lands, and separated from your tribal ways, though some of you still have your traditional culture. Some of you are establishing Aboriginal communities in the towns and cities. For others, there is still no real place for campfires and kinship observances except on the fringes of country towns. There, work is hard to find, and education in a different cultural background is difficult. *The discrimination caused by racism is a daily experience.*
>
> ADDRESS TO THE ABORIGINES AND TORRES STRAIT ISLANDERS,
> ALICE SPRINGS, AUSTRALIA, NOVEMBER 29, 1986.

Cultural Homelessness, Depression, and Suicide

When ecological violence is perpetrated on indigenous people and their homelands, the damage is not only physical, but spiritual, as it erodes their personal and cultural identity, way of living, values, and community bonds. The violence that began in the centuries of land-grabbing and wars has continued through corporate and government takeovers of indigenous lands and natural resources. Endemic poverty, dreadful cultural diseases, and public-health issues have resulted from land and resources being stolen, combined with environmental and social racism and violence. All have been accentuated by the burden of the genetic and addictive disease of alcoholism. A 2013 public health study of native peoples of Canada, the United States, New Zealand, and Australia found that high levels of alcoholism and drug use are exacerbating the risk of mental-health issues, especially depression and suicide:

…their higher levels of social disadvantage increases their exposure to stressful life events, such as unemployment, homelessness, incarceration and family problems, that, in turn, have been shown to increase one's risk of suicide. Indigenous peoples of Australia, New Zealand, Canada and the United States are also at an increased risk of suicidal behaviour due to factors embedded in their historical experiences, including loss of land and culture, transgenerational trauma, grief and loss, racism and social exclusion.

ANTON C. CLIFFORD, INSTITUTE FOR URBAN INDIGENOUS HEALTH;
CHRISTOPHER M. DORAN, HUNTER MEDICAL RESEARCH INSTITUTE; AND
KOMLA TSEY, CAIRNS INSTITUTE, JAMES COOK UNIVERSITY; AUSTRALIA

In Australia, 40% of aboriginal people are homeless, and the youth suicide rate among aboriginal people is the highest in the world. The UN's 2011 "State of the World's Indigenous Peoples" report detailed further shocking statistics:

In the United States, a Native American is 600 times more likely to contract tuberculosis *and 62% more likely to commit suicide than the general population.* In Australia, an indigenous child can expect to die 20 years earlier than his nonnative compatriot. The life expectancy gap is also 20 years in Nepal, while in Guatemala it is 13 years and in New Zealand it is 11. In parts of Ecuador, indigenous people have 30 times greater risk of throat cancer than the national average. *Suicide rates of indigenous peoples, particularly among youth, are considerably higher in many countries, for example, up to 11 times the national average for the Inuit in Canada.*

Suicide rates among indigenous youth are so high worldwide that multiple conferences have been called everywhere from Brazil to Micronesia to try to address the varied situations.

This is the other side of the spectrum of the culture of waste—not one resulting from materialism but exploitation, loss, and poverty. The more the mainstream culture looks down on and dismisses native cultures, the easier it is for people to treat them as dispensable, and the cycle continues. Pope John Paul II called upon native and nonnative faith communities to work together on the shared societal problems that hit the native communities particularly hard because of their fewer resources and harsh history:

> At the same time I call upon your native Catholic communities *to work together to share their faith and their gifts,* to work together on behalf of all your peoples. There is much to be done in solving common problems of unemployment, inadequate health care, alcoholism and chemical dependency. You have endured much over hundreds of years and your difficulties are not yet at an end. Continue taking steps towards true human progress and towards reconciliation within your families and your communities, and among your tribes and nations....
>
> ADDRESS TO THE NATIONAL TEKAKWITHA CONFERENCE,
> PHOENIX, ARIZONA, SEPTEMBER 14, 1987

The world, in turning toward a new value of caring for the earth, must also turn toward caring for its youth—*all* youth—and future generations. John Paul II said we must stand with native peoples, honoring and working alongside them as one human family so we might all survive together with new attitudes and a new relationship to the earth.

> You have kept your sense of brotherhood. If you stay closely united, you are like a tree standing in the middle of a bush-fire sweeping through the timber. The leaves are scorched and the tough bark is scarred and burned; but inside the tree the sap is still flowing, and under the ground the roots are still strong.
>
> Like that tree you have endured the flames, and you still have the power to be reborn. The time for this rebirth is now!...
>
> ADDRESS TO THE ABORIGINES AND TORRES STRAIT ISLANDERS,
> ALICE SPRINGS, AUSTRALIA, NOVEMBER 29, 1986

Theology, philosophy, and science all speak of a harmonious universe,
of a "cosmos" endowed with its own integrity,
its own internal, dynamic balance. This order must be respected.
The human race is called to explore this order,
to examine it with due care,
and to make use of it while safeguarding its integrity.

"PEACE WITH GOD THE CREATOR,
PEACE WITH ALL CREATION," WORLD DAY OF PEACE, JANUARY 1, 1990

10. Loss of Biodiversity, Endangered Species, and Interdependence

A reporter once asked the scientist J.B.S. Haldane (reportedly an agnostic) what he had learned from a lifetime studying natural history and genetics. Haldane responded, "God must have an inordinate fondness for beetles." Beetles happen to be one of the most diverse orders of animals on earth, accounting for about one third of living species.

Why so many beetles? one might ask. *What good are they? What would it matter if we lost some of them?*

Pope John Paul II answered questions like this in this way:

> Destruction to any part of creation harms the whole. The ensemble of creatures constitutes the universe. In its totality, as well as its parts, the visible and invisible cosmos reflects eternal Wisdom and expresses the inexhaustible love of the Creator

GENERAL AUDIENCE, MARCH 12, 1986

Creation has provided all the species and "parts," therefore they are needed even if we can't see the greater purpose each serves. It's not for us to judge, as God has already judged and blessed each piece, and the whole, as *good.*

But let's look at things ecologically. Beetles, for example, are generally scavengers and do the cleanup work for the planet, eating and recycling food waste and dead matter. So they're fairly important in keeping the flow of nutrients moving into the soils. They're also food for other species. Insects are more varied than the rest of life on earth because their habitats are smaller, and their job specifications—what they eat and how they exist— are more diverse. The functional niches they fill are like individual steps on intersecting assembly lines. Because of their size and variety, it may seem that we could easily do without a few of them. But we would be wrong. Ecologist Aldo Leopold once said, "To keep every cog and wheel is the first precaution in intelligent tinkering" ("THE ROUND RIVER," *A SAND COUNTY ALMANAC: WITH ESSAYS ON CONSERVATION FROM ROUND RIVER*).

Here's why: During an Amazon winter, the only trees that flower and bear fruit are fig trees. As you can imagine, a lot of animals—toucans, bats, monkeys, and many others—are pretty desperate for that fruit. Jaguars and other predators then move in to hunt the monkeys and other fig eaters. What is interesting about this food web is that it all depends on a single kind of insect—a fig wasp. There are about nine hundred species of figs, and each one has its own specific species of fig wasp that evolved along with it, living its life in and among the flowers of that particular type of fig plant. Along the way, the wasps perform the crucial role of pollinator for their sister species. Without this pollination by the particular wasps, the fig crops would fail, and those who feast on figs in the winter months would be vulnerable to starvation. Elements harmful to the fig wasps, such as pesticides or diseases, endanger the entire ecosystem.

This may seem like an academic example (unless of course you are a fig farmer, fig-loving monkey, or a jaguar), but it's fundamental. Many insects that seem like "throwaways" to the lay person are essential in the networks of life. Currently, due to the prevalence of agricultural pesticides and imported mites and diseases, pollinating bees in North America are under great threat. Add stresses from the increased temperatures from global warming and you have an ecological mess. Native bee populations are plummeting, and bee-keepers are finding it harder and harder to maintain their domesticated hives over the winter. What will happen to the breadbaskets around the world if the populations of pollinators completely collapse, which is the path they seem to be on? Could we pollinate enough crops by technology or by hand to make up for these insects? We couldn't. A collapse in pollinators would mean famine.

As has been noted, we are losing a staggering number of species, 25,000 to 50,000 a year, many of which have not even been identified. Some scientists are saying that we are driving the planet to its biggest mass extinction since the dinosaurs died out 65 million years ago. This is occurring because of climate change, overhunting, poaching, habitat loss, invasive species, and pollution.

One might say, well, we must have an awful lot of extra species around to be losing this many a year without our noticing it. Scientists have been noticing it, but the general public hasn't noticed it *yet*. No one knows what the tipping point will be, the point at which the biological systems under strain will simply sputter, cough, and break down.

Pope John Paul II called for a shift in vision and understanding—a shift that acknowledges the value and interconnectivity of all of life. "One must take into account the nature of each being and of its mutual connection in an ordered system, which is precisely the 'cosmos'" (ENCYCLICAL *SOLICITUDO REI SOCIALIS*, 1988).

Is It Really Endangered Species Versus Jobs?

A loss in biodiversity brings about losses in food webs, fertility, oxygen and humidity, sources of foods and medicines, wildlife, protection against agricultural diseases, and so much more. Though the indirect damage is inestimable, scientists and economists warn that the direct monetary losses mount exponentially when ecosystem fertility, scenic value, water and air quality, pollination, and many other environmental necessities are lost. So the fight to save endangered species around the world is emotional, economical, and essential. Pope John Paul II stated:

> Delicate ecological balances are upset by the uncontrolled destruction of animal and plant life or by a reckless exploitation of natural

resources. It should be pointed out that all of this, even if carried out in the name of progress and well-being, is ultimately to humankind's disadvantage.

WORLD DAY OF PEACE 1990 MESSAGE "PEACE WITH GOD THE
CREATOR, PEACE WITH ALL CREATION," JANUARY 1, 1990

Consider how often, however, we hear the protection of endangered species and habitats pitted up against "jobs" in public debates, as if these things were diametrically opposed *instead of intimately connected.* It's often easier for people to blame those trying to protect wildlife for job losses or a downturn in an economy instead of the way our businesses and stocks are structured for short-term profits rather than long-term husbandry. Sustainable practices often require more people power—*more* jobs—because the processes involved are slower, more individualized, and more methodical (as in sustainable forestry versus clear-cutting). Yet profit margins and large benefits packages to CEOs, top management, and board members drain money from the bottom line so that the proper amount of personnel and livable wages can't be budgeted in to support sustainable practices.

Pope John Paul II recognized greed, selfishness, short-sightedness, and ignorance as some of the stumbling blocks to living harmoniously with the rest of nature. He declared:

> Today, the dramatic threat of ecological breakdown is teaching us the extent to which greed and selfishness—both individual and collective—are contrary to the order of creation, an order that is characterized by mutual interdependence.

WORLD DAY OF PEACE 1989 MESSAGE,
"TO BUILD PEACE, RESPECT MINORITIES," DECEMBER 8, 1988

He counseled that we must be thinking beyond short-term profits to the effects our actions can have on species and systems, the poor, middle-class workers, and generations to come: "No intervention in an area of the ecosystem can neglect weighing the consequences in other areas and...the effects it will have on the well-being of future generations" (MESSAGE FOR THE 23RD WORLD DAY OF TOURISM, JUNE 24, 2002).

To Diminish the Diversity of the Earth
Is to Diminish Human Life

Thankfully, in our day, there is no lack of scientists, wildlife managers and activists, conservationists, and dedicated organizations working to elucidate some of these ecological effects and what is necessary to protect endangered species, habitats, and systems. What *is* lacking is an understanding among Roman Catholics and other people of faith about the inherent connections built into the earth, and that caring for all nature is part of faith. John Paul praised the scientists for unveiling the connections:

> It is to the undeniable credit of scientists that the value of biodiversity of tropical ecosystems is coming to be more understood and appreciated.... The extent of the depletion of the earth's biodiversity is, indeed, a very serious problem. Even the quality of human life, because of its dynamic dependence on the interaction of other species, is being impoverished.
>
> ADDRESS TO PONTIFICAL ACADEMY OF SCIENCES STUDY WEEK ON MAN AND HIS ENVIRONMENT, TROPICAL FORESTS, AND THE CONSERVATION OF SPECIES, MAY 18, 1990

Aldo Leopold explained ecological interdependence and diversity in simple relational terms:

> Harmony with the earth is like harmony with a friend: you cannot cherish his right hand and chop off his left…you cannot love game and hate predators; you cannot conserve the waters and waste the ranges; you cannot build the forest and mine the farm. The land is one organism.
>
> "THE ROUND RIVER," *A SAND COUNTY ALMANAC: WITH ESSAYS ON CONSERVATION FROM ROUND RIVER*

It almost sounds like the pope read Leopold when John Paul II said to California rural workers in 1987, "We cannot say we love the land and then take steps to destroy it for use by future generations" (HOMILY, MASS FOR RURAL WORKERS IN LAGUNA SECA, MONTEREY PENINSULA, CALIFORNIA, SEPTEMBER 17, 1987).

Global Solidarity to Save Species and Biodiversity

Recognizing the value of diversity, international leaders signed the Convention on Biodiversity at the 1992 United Nations Earth Summit. (President Clinton signed it, but the U.S. Congress has yet to ratify it. Sadly, the United States is the only UN member that has still not officially agreed to the treaty.) The agreement has three core principles: conservation of biological diversity; sustainable use of its components; and fair and equitable sharing of benefits arising from genetic resources.

These principles also apply to the protection of marine biodiversity. They were reaffirmed in the UN Millennium Development Goals and then strengthened by the Cartegena Protocols of Biosafety of 2003, which have been ratified by 166 countries so far. (The United States has not acted on this yet either.)

Bringing the principles of this convention into action requires cultural value shifts around the world, especially in developed countries where corporations yield such power over politics. The pope pinpointed this lack of moral fortitude and action the following year, at the 2004 World Food Day, whose theme was biodiversity:

> Many obstacles today stand in the way of international action to conserve biodiversity. Despite the existence of increasingly effective regulations, other interests seem to upset the just balance between the sovereignty of States over the resources in their territory and the ability of individuals and communities to retain or manage these resources in terms of real need.
>
> Therefore, international cooperation must also be based on the principle that claims of sovereignty over genetic resources in ecosystems cannot become exclusive nor a cause of conflict; they must be exercised in accordance with the natural rules of humanity that govern coexistence among the different peoples that make up the human family.
>
> MESSAGE TO JACQUES DIOUF ON 24ND WORLD FOOD DAY, OCTOBER 15, 2004

The Pope's Position Based on Science and Faith

Though an espouser of *both* evolution and divine intelligence, Pope John Paul II did not feel that any specific faith or scientific position was necessary to appreciate the beauty of the interconnected universe. The saints and Catholic teachings have always rejoiced in the sacredness of all of life as expressions of the Divine, of God at work in the world. In application of this, the pope presented this philosophical equation to Catholics, other Christians, and all who believe in a good Creator: if such a wondrous and mysterious and powerful God spurred the coming into being of the many different species of life and the web of biological diversity and pronounced them *good*, then they *are* inherently good, and as humans entrusted with creation we *must* protect them:

> Every kind of life should be respected, fostered, and indeed loved, as a creation of the Lord God, who created everything 'good.'
>
> PONTIFICAL ACADEMY OF SCIENCES STUDY WEEK ON MAN AND HIS ENVIRONMENT, MAY 18, 1990

Who cannot but be deeply concerned by the prospect of the already existing and ever expanding danger from pollution and other side effects of the production and use of chemicals?

ADDRESS TO THE PONTIFICAL ACADEMY OF SCIENCES, WORKSHOP ON CHEMICAL HAZARDS IN DEVELOPING COUNTRIES, OCTOBER 22, 1993

11. Chemical and Industrial Pollution

On December 2, 1984, a pesticide plant in Bhopal, India, run by a subsidiary of the Union Carbide Corporation (UCC), exploded in a gas leakage disaster. Thousands died in the immediate firestorm of chemicals. The plant had been located in an area of shantytowns, whose poor residents did not have the money or political power to fight it. A number of smaller leakages had occurred prior to the date of the disaster. From 1979 to 1984, plant authorities had been issued warnings by the workers' union, an American expert, and others that things were unsafe. But all safety and maintenance issues were ignored to cut costs.

In addition, the company had been storing hazardous waste in the area, and by 1982, one hundred water wells used by the local community had already been destroyed by pollutants. But no one listened to the residents' complaints. Soils were ruined, with high levels of toxins showing up in the local vegetables and breast milk of mothers. No press coverage, no pressure for cleanup, no justice followed these public health disasters. The damage for which no one felt responsible would have continued at Bhopal w*ithout any problems for the corporation*—except an even worse disaster overtook them.

This pattern of problems and warnings being ignored to cut costs was the same in those leading up to the 2013 collapse of the clothing sweatshop in Bangladesh, which killed over 1,100 workers, or the British Petroleum *Deepwater Horizon* fiasco in the Gulf of Mexico. There was nothing "accidental" about these disasters—they were merely an intensified version of what was already happening, the logical consequences of cost cutting over safety, and a lack of ethical corporate leadership.

When the chemical gas exploded in the Bhopal factory, the initial death count was 2,259, which Indian authorities later raised to 3,787 immediate deaths, 7,000 to 8,000 over the first two weeks, and a further 7,000 to 8,000 over the ensuing years, leading to an official death toll of 15,000. A dense, toxic gas cloud of 40 tons of methyl isocyanate and numerous other toxic fumes, worse than a chemical weapon release, covered thirty-six wards of the capitol city, affecting from 500,000 to 700,000 people, including approximately 200,000 children. The fumes caused blindness, suffocation, shutting down of organs, and later, cancers, respiratory and neural diseases. Beyond the enormous death toll, the official 2006 government tally listed 58,125 injured, with 3,900 permanently disabled, and 38,478 temporarily disabled. Stillbirths in Bhopal also increased by approximately 300% and prenatal deaths by 200%.

Hunger multiplied exponentially, as 2,000 livestock died and were left to bloat, causing noxious smells, flies, and disease. Mass cremations of people and animals added to the air pollution, while water sources were polluted by both the gases and the decaying bodies. Any fish remaining were too toxic to eat, as were vegetables and other foodstuffs. The trees and other vegetation were acid burned and lost their leaves.

Pope John Paul II Calls for a Moral Response and Preventative Vision

In the wake of the Bhopal disaster and the *Exxon Valdez* oil spill a year later, Pope John Paul II told the United Nations that businesses and nations *must be made to take on the responsibility for past tragedies and that they must also internalize a moral element into their decision-making at all levels* to plan ahead and find less toxic ways to make products, in order to protect workers and the environment. He wanted them to place human faces on the damages, and not look at the costs as merely lines on their accounting sheets—to respect life, not money.

He told the UN Food and Agricultural Organization:

> Economic activity carries with it the obligation to use the goods of nature reasonably. But it also involves the grave moral obligation both

to repair damage already inflicted on nature and to prevent any negative effects that may later arise.

A more careful control of possible consequences on the natural environment is required in the wake of industrialization, especially in regard to toxic residue, and in those areas marked by an excessive use of chemicals in agriculture.

ADDRESS TO THE XXV SESSION OF THE CONFERENCE OF THE UN
FOOD AND AGRICULTURAL ORGANIZATION, NOVEMBER 16, 1989

The pope asked the world to care and assist for Bhopal's suffering survivors and prayed "that they will experience the fullness of fraternal solidarity of which they have need" (LITURGY OF THE WORD AT THE AIRPORT OF MANGALORE, INDIA, FEBRUARY 6, 1986).

The Suffering Continues. So Does the Struggle.

The Bhopal disaster may seem like old news, but the people in Bhopal still face painful challenges daily, coping with what was left behind. A memorial reads: "The suffering continues. So does the struggle." On the day after the disaster, MIT-trained engineer Satinath Sarangi arrived on the scene and, seeing the extent of suffering, devoted himself to advocacy for the victims. Though Union Carbide paid out the assigned damage amounts (the equivalent of about $2,000) to a specified number of individuals, built a hospital, and did some remediation cleanup, the company in no way paid out the full damages or could possibly restore fully all that had been polluted.

The president of U.S. Union Carbide, Warren Anderson, eluded arrest for any responsibility, though charges in India have been pending against him for decades; he was labeled an "absconder." Finally in 2010, nearly thirty years after the accident, seven corporate leaders of the Indian subsidiary went on trial in India. They were held accountable *only for neglect* and charged with two years in jail and a 100,000 rupee ($2,125) fine.

"They have treated it like a traffic accident," said Sarangi, decrying the Indian Supreme Court ruling as another disaster of justice, layered on the first. Rashida Bee, the president of the Gas Women's Workers group (one of the unions representing Bhopal workers), said "justice will be done in

Bhopal *only* if individuals and corporations responsible are punished in an exemplary manner" (AGENCE FRANCE PRESS, JUNE 7, 2010).

Corporate Irresponsibility and Lack of Personal Accountability

The Bhopal explosion, the *Exxon Valdez* oil spill, the BP disaster in the Gulf of Mexico, the 2013 sweatshop collapse in Bangladesh...Not much has changed. Even as oil companies are proposing thousand-mile pipelines such as the United States-Canada Keystone and Enbridge expansion projects, they are not reporting the many leakages and spillages along existing lines that are already doing destruction. This same story is repeated in industry after industry. Who is responsible for the chemical and industrial pollution caused? No one personally.

This is precisely the cold evil and immoral structure of sin that Pope John Paul II spoke about. In most cases, only the business or corporate legal structure is held responsible (if anyone finds out about the problems and the company is forced to be accountable), not the decision-makers in the board-rooms and executive suites. When profits are the only driving motivation, damage to the environment and people result, for now and for generations to come. The pope made this very clear:

> Destruction of the environment highlights consequences of decisions made by private interests that do not weigh the real conditions of human dignity. One finds prevalent an unbridled desire to accumulate personal wealth that prevents people from hearing the alarming cry of poverty of entire peoples.... The selfish quest for their own good fortune induces people to disregard the legitimate expectations of present and future generations.

MESSAGE FOR THE 23RD WORLD DAY OF TOURISM, JUNE 24, 2002

History of American Corporate Accountability

Back in 1816, Thomas Jefferson warned about the lack of corporate accountability and its dangers to the new American nation:

The country is headed toward a single and splendid government of an aristocracy founded on banking institutions and moneyed incorporations and if this tendency continues, it will be the end of freedom and democracy, the few will be ruling....I hope we shall crush...in its birth the aristocracy of our moneyed corporations, which dare already to challenge our government to a trial of strength and bid defiance to the laws of our country.

LETTER TO GEORGE LOGAN, NOVEMBER 12, 1816,
THOMAS JEFFERSON RETIREMENT SERIES DIGITAL LIBRARY.

He warned lawmakers never to let corporations become "immortal persons" in the eyes of the law, as they had become in Europe, wreaking so much havoc on the freedoms of everyday people. The early states of the nation took his counsel to heart, along with similar earlier advice from Benjamin Franklin and other Founding Fathers. However, the states also recognized the usefulness of small, restricted corporations to accomplish large projects efficiently. So the states provided limited-time charters of license to corporations that were organized to accomplish specific activities perceived as being for the common good, such as road or canal building.

If the corporation or its officers *violated any laws, or if the company charter was violated, the corporation was forced to dissolve.* Officers, managers, and owners *were held liable for any crimes* they committed under the auspices of the corporation, and if the corporation sought to infringe on any rights of the public or harm it in any way, the charter would be terminated. Finally, *corporations were banned from contributing to political campaigns or endeavors, or seeking in any way to influence politics.* Over the centuries, such laws were modified, against all of Jefferson's warnings, and now U.S. corporations have "personhood" status with little criminal accountability for decision-makers and great scope for lobbying and political influence.

Organizations such as Move to Amend have started a grassroots campaign in America and other places to take back personhood status from corporations and businesses, to hold corporate decision-makers responsible for the policies or choices they make in the name of the company, and to stop corporations from intense lobbying and contributions to political campaigns. Some corporations have set up internal ethical and ecological standards for action, a good first step, but they are not measured by outsiders nor held as legally binding.

Laws of a global, free market have often been used to circumvent, overrule, or abuse local laws, mores, environments, and people. Irresponsibility has become systematized internationally in the name of profit and shareholder value. Many have stigmatized any government restrictions or regulations, especially in the area of environmental protection, as "interfering with the market," which is tantamount to a secular sin in their minds. This way of thinking deeply disturbed Pope John Paul II.

> If globalization is ruled merely by the laws of the market applied to suit the powerful, *the consequences cannot but be negative.* They are, for example, the absolutizing of the economy, unemployment, the reduction and deterioration of public services, the destruction of the environment and natural resources, the growing distance between rich and poor, unfair competition which puts the poor nations in a situation of ever increasing inferiority.
>
> POST-SYNODAL APOSTOLIC EXHORTATION, *ECCLESIA IN AMERICA*, 1999

Pope Francis has been even more adamant. In his apostolic letter of July 11, 2013, he stated: "In our times, the common good is increasingly threatened by transnational organized crime, the improper use of the markets and of the economy, as well as by terrorism." He put his official stamp on an act to hold all officials and criminals accountable across international borders and within the Vatican state. The new "golden calf" of global markets, he feels, needs melting down into new more service-oriented forms for sustainable job creation and the common good.

The Multiplying Profusion of Chemicals and Their Effects

The number of chemicals presently in circulation in the world is astounding: over 7 million, with more than 80,000 in common use. The U.S. Environmental Protection Agency has 84,000 chemical substances listed in its inventory of chemicals in use in agriculture, manufacturing, cosmetics, food, and other industries. From 500 to 1,000 are added each year in the

United States alone, with little to no testing for human and wildlife toxicity. And there are no ways to test how living beings, human and wildlife, can handle the combinations of the various chemical compounds subtly affecting us all at once.

The abundant toxins in the environment are invading our very bodies. Pesticides coat supermarket produce and accumulate in people's cells, with most of us having high rates of various chemicals in our fat and tissues. Chemicals in the environment are linked to many illnesses, including cancer, and changes in the hormonal responses of children, women, and wildlife. Mercury causes birth defects and poisoning, as does nuclear waste. Children have rising rates of asthma due to chemicals and particulates in the air. Researchers are exploring the connections between the rise in chemical pollution and rising rates of allergies, arthritis, and other autoimmune diseases, as well as infertility and developmental problems. David Bellinger of the Harvard School of Public Health says "There have been several prospective studies that show a relationship between pesticide levels in pregnant women and lower IQ levels in children" (DR. SUSHRUT JANGI, JULY 24, 2013, BOSTON.COM). The list goes on.

The 2012 "Global Chemicals Outlook," a UN Environment Program report, painted a fairly depressing international picture. It stated that chemical poisoning results in approximately 964,000 deaths annually (1.6% of all deaths). It also noted that there are more than 140,000 chemicals at use regularly in the world market, and only a fraction have been tested sufficiently for health safety and environmental effects. Yet they appear in all sorts of consumer items, including children's toys.

Pope John Paul II called attention to the damage caused by chemicals in 1990, when he called the ecological crisis a moral crisis:

> Unfortunately, it is now clear that the application of these discoveries in the fields of industry and agriculture have produced harmful long-term effects. This has led to the painful realization that we cannot interfere in one area of the ecosystem without paying due attention both to the consequences of such interference in other areas and to the well-being of future generations.
>
> WORLD DAY OF PEACE MESSAGE "PEACE WITH GOD THE CREATOR, PEACE WITH ALL CREATION," JANUARY 1, 1990

The Blacksmith Institute, an award-winning international not-for-profit organization that works with communities to clean up toxic sites and prevent pollution, issues an annual report on the most toxic site-based sources. In 2012, the top fifteen worldwide sources (some overlapping) emitting known toxic substances, including lead, mercury, cadmium, arsenic, cyanide, chromium, heavy metals, benzene, formaldehyde, toluene, vinyl chloride, and chloroform, and many others, were: Lead-acid batteries and improper battery recycling facilities; lead smelting; surface and underground mining and ore processing; tanneries; industrial and municipal dump sites; artisanal gold mining; product and food manufacturing; chemical and petrochemical manufacturing (including fertilizers); dyeing and painting of textiles and products; electronic manufacturing and waste; heavy metals facilities; pesticide manufacturing and storage; and uranium processing.

This list focuses on specifically on toxins concentrated at base operations, not the chemicals that flow through the system in the foods we eat and the products we use, or the pollution from power plants. The Blacksmith Institute noted that agricultural chemicals, specifically pesticides, have been rated in the past as one of the highest polluters internationally, and it feels confident that they still are, but it admits that it is very difficult to get accurate statistics on the health consequences of the pesticide industry because the use of agricultural chemicals is so endemic.

It isn't just the pesticides either. For instance, the U.S. Geological Survey estimated that in May 2013 alone 153,000 metric tons of chemical nutrients, mostly from agricultural phosphorus and nitrogen, flowed down the Mississippi and Atchafalaya Rivers to the northern Gulf of Mexico. These nutrients caused algae to bloom profusely, absorbing the oxygen in the ocean water and adding to the Gulf's "dead zone." This dead zone, extending into the ocean from the Mississippi Delta, grows yearly by approximately the surface size of Connecticut.

As for mining, surface extraction methods, such as mountaintop removal, are now the cheap, favored forms. These processes start with stripping the land of all forest and vegetation, then peeling away the earth's surface soils and crust. The "fillings" are dropped into valley slag piles, which wash into streams. The rock underneath the crust is then dynamited, filling the air with toxic particulates. The actual drilling uses millions of gallons of water and leaves leaching waste ponds. The result is miles and miles of stripped mountains and chemically polluted watersheds and human and wild communities. (If you are a landowner and the government has sold or leased the area's mineral or exploratory rights, an extraction company has the power to destroy your property in order to reach what is encased underneath. Their legal rights trump yours, so you can lose your home and way of life with minimal recourse and have to settle for the compensation offered.)

Chemicals in Developed Versus Undeveloped Countries

Developed countries often have established at least minimum regulations in most industries for worker safety, emissions, storage, and dumping, though they are not generally sufficient to protect the ecological systems upon which they depend. However, less industrially advanced countries have fewer tools of protection in place, so multinational corporations like Union Carbide (now known as Dow Chemical) and others often locate chemical-intensive plants, mining, and oil drilling in nations or states where there are fewer workers' rights, safety, clean air, soil, and water restrictions, oversight, and enforcement. National governments, seeking greater tax revenue and jobs, often offer multinational corporations within their borders little oversight or pressure to protect the people or environment. The Blacksmith Institute's list of toxic industries highlights this, with the majority of the worst industry sites in more poverty-struck countries.

Pope John Paul II didn't accept these practices as necessary simply because they were endemic. He believed we can change, if we have the religious, moral, and political will:

> The legitimate development to which every country aspires cannot
> be pursued irresponsibly, at the cost of the natural environment. Entire
> regions of the globe are threatened by excessive exploitation of natural
> resources and by inadequately controlled pollution. Nations and indi-
> viduals have a moral duty to protect the common patrimony of animal
> and plant life, and to avoid contaminating land, sea and air…
>
> ADDRESS TO THE AMBASSADOR OF MAURITIUS, VATICAN CITY, APRIL 24, 1997

Children Are Hurt the Most

As would be expected, whether in a developed or developing country, dumps, toxic storage sites, mining, drilling, and chemical plants tend to be located near people with the least affluence and political voice. Children suffer the most, since their bodies are smaller and the toxic doses are thus proportionally larger and more damaging to their vulnerable, still-developing bodies.

Recent cases emphasize the dire situation. In northwest Nigeria, over 400 children under the age of five were slowly poisoned and killed by local gold mining operations, which gave off lead dust. Few in the nearby farm communities had any idea what was sickening their young children. In addition, over 2,000 children were left permanently disabled with brain damage, paralysis, and deformities.

In India, Pepsi and Coke products have been found to have 30 to 36% more pesticides in them than the recommended standards, and New Delhi bottled water had 40 times the acceptable amount. Farmers with little training or knowledge of the dangers commonly spray pesticides banned in the United States and Europe on their rice and grains, often without masks, and use the empty containers for food storage. Cancer rates are at epidemic levels. In July 2103, 23 children in northern India died of pesticide poisoning after eating their school lunch, which was linked to pesticide-tainted cooking oil stored in a toxic container.

Pope John Paul II pointed out the sheer immoral nature of these double standards for different countries and the need for higher standards worldwide. *What is unsafe in one country is unsafe in any country*, and multinationals from affluent countries should not make money in other countries from products or processing considered unsafe in their home countries. This is exporting evil. It is another form of imperialism that preys upon the poor and vulnerable. He said:

> ...in the developing countries, where most chemical hazards have their origin in the import of chemical substances and technologies, a lack of expertise and of necessary infrastructures often renders efficient control difficult or impossible. Very few countries, in fact, have a specific legislation regulating the handling and use of toxic chemicals....

It is a serious abuse and an offense against human solidarity when industrial enterprises in richer countries profit from the economic and legislative weakness of poorer countries by locating production plants or accumulating waste that will have a degrading effect on the environment and on people's health.

ADDRESS TO THE PONTIFICAL ACADEMY OF SCIENCES, WORKSHOP
ON CHEMICAL HAZARDS IN DEVELOPING COUNTRIES, OCTOBER 22, 1993

In contrast, when international businesses and their leaders look at themselves as *having a vocation of service that also makes money*, a multinational corporation with great access to wealth, experts, and technologies can accomplish so much to protect those who work for them and their environments. They have the power to attain *far more than the local national norms* in order not only to bring jobs but to provide local people the same quality of life and protections the business people enjoy in their own home countries. In this way, they are exporting, along with their products and jobs, goodness instead of cold evil—positive changes for poverty alleviation and pollution solutions. As the pope quoted the Epistle to the Romans, "Be not overcome by evil, but overcome evil with good" (WORLD DAY OF PEACE MESSAGE, 2005).

Obligations of Protection and Care

As much as Pope John Paul II put a heavy moral burden on corporations and industry, he laid equally hefty responsibilities on the international community and individual nations to oversee them together and set goals for a better future. The countries where multinational corporations are headquartered should require that their national standards (or those of the nations where the business is being conducted, if those standards are higher) be upheld in commercial and industrial ventures. The pope hammered this point home: "It would be difficult to overstate the weight of the moral duty incumbent on developing countries in their efforts to solve their chemical pollution and health hazard problems" (ADDRESS TO THE PONTIFICAL ACADEMY OF SCIENCES, WORKSHOP ON CHEMICAL HAZARDS IN DEVELOPING COUNTRIES, 1993).

Human beings, by nature, need systems of checks and balances to guide and limit their behaviors, which is why the markets need oversight, and why a certain amount of government and international restrictions and enforcement will always be necessary for the common good. Pope John Paul II said "The destruction of the environment caused by industry and industrial products must be reduced in accordance with precise plans and undertakings, also at the international level" (ADDRESS TO THE PONTIFICAL ACADEMY OF SCIENCES, STUDY WEEK ON RESOURCES AND POPULATION, NOVEMBER 22, 1991).

In 2002, at the UN World Summit on Sustainable Development, also known as the Johannesburg Earth Summit, the international community set the goal of minimizing the impact of chemicals on human health and the environment by 2020. The "Global Chemical Outlook" report offered twenty-six strategies to help the world reach its goals, but they look very difficult to attain.

With a master's degree in chemistry, Pope Francis may be able to speak with even greater knowledge and moral urgency on this issue, to persuade the international community to take its goals seriously by banning specific hazardous chemicals and practices and by enforcing higher standards for production and use of other chemicals—and by making corporate officers and business leaders accountable for their actions.

To address these issues, a new collaborative body that includes the banking sector has grown out of the UN negotiations for the Minamata Convention on Mercury, which was convened to curb use of this hazardous substance and signed by participating nations in October 2013. The Global Alliance on Health and Pollution (GAPH) aims to reduce hazardous pollution and help low- to middle-income nations clean up toxic hot spots and prevent future toxic pollution. Members include the World Bank, the Asian Development Bank, and the UN Industrial Development Organization. The Blacksmith Institute is the guiding consultant and association secretary. The formation of this powerful partnership is another step toward international cooperation to address the pandemic proliferation of chemicals.

To aid in oversight and transparency, everyday citizens around the world are now armed with a new nonviolent weapon to fight chemical and industrial pollution: a cell phone. A consulting firm called Nexleaf works with community organizations around the globe to assist local people in reporting incidents of air, soil and water pollution, spills and leaks, via cell phone photos, videos, texts, and tweets. These can be sent to an open source webpage as well as communicated to media, officials, nonprofit organizations and NGOs, parishes and diocesan offices, other activists, and blogs. The purpose is to bring the light of testimony to the darkness of coverups. Neighborhood and village watches can be set up to keep an eye on worrisome plants or industrial operations, with cell phones in hand to document infractions and make them public. Chemical detecting drones are also useful for legislative oversight.

The Toxic Home Front

However, industries and corporations are not the only ones to hold accountable for chemical pollution. We all need to watch ourselves. Consider how many unpronounceable food additives are in the processed food products we gobble daily and feed to our children, increasing the market demand for more chemical processing. Mosquito spraying settles over cities. The lawns of homes, churches, parks, businesses, and municipal buildings are doused with fertilizers and pesticides. When you walk through many U.S. cities in the spring, instead of smelling the scent of the lilacs and cherry blossoms, you often smell herbicides being sprayed on the winds or rising up out of treated lawns. Little signs pop up instead of dandelions: "Please keep off the lawn as it is dangerous to the health of animals and children." What is a lawn for if not to enjoy it with your kids and pets? What of the wildlife who can't read signs?

In addition, the products we buy and the garbage we toss create toxic waste as they decompose. Consumers have the power to change this by not buying products that contain toxic chemicals, not eating foods that have ingredients we don't recognize, and not using chemicals on our lawns and gardens. We can stop buying products from corporations that have double standards for safety and environmental actions at home versus abroad. Like groups such as the Moms Clean Air Force and so many others, we can pressure leaders for more restrictive legislation and vote for politicians who respect and protect life in all its forms, recognizing the connections between the environment and harm to children and the unborn as well as to other species.

Pope John Paul II urged Roman Catholics and other Christians and people of faith and goodwill to stand up in their homes, boardrooms, businesses, congregations, and political arenas to push for a full respect for life that sees these connections. Nations, scientists, corporations, shareholders, and everyday citizens must retool their relationship with people and nature into one of duty, self-control, justice, sacrifice, service, and love that is acted out to protect all of life. In his address to the Pontifical Academy of Sciences workshop, he summed up his thoughts on the subject:

> Respect for the natural environment and the correct and modulated use of creation's resources are a part of each individual's moral obligations....

> ADDRESS TO THE PONTIFICAL ACADEMY OF SCIENCES, WORKSHOP ON
> CHEMICAL HAZARDS IN DEVELOPING COUNTRIES, OCTOBER 22, 1993

The poor man of Assisi gives us striking witness that when we are at peace with God we are better able to devote ourselves to building up that peace with all creation, which is inseparable from peace among all peoples.

WORLD DAY OF PEACE 1990 MESSAGE "PEACE WITH GOD THE CREATOR, PEACE WITH ALL CREATION," JANUARY 1, 1990

12. War and Peace

Twenty years after the war in Vietnam, a girl is sent out to hoe a fallow field. She hits a landmine and her leg is blown off. She is rushed by cart to get medical help, and now she will need financial assistance to get an artificial limb. Her injury puts even greater strain on her hungry household, and she may find it difficult to find a husband.

During the war, the American armies drained the wetlands of the Mekong Delta and used Agent Orange and napalm to defoliate the cajeput and mangrove forests. The leaves disappeared, and the wildlife went with them. The farming of rice was made even more difficult (and toxic) as much of the water was gone, and the missing forests couldn't hold down the moisture. The ground and remaining water were poisoned by the chemicals, making birth defects and miscarriages common occurrences in the ensuing years.

The Vietnamese have a saying: "Birds only stay in good lands." The eastern sarus crane, an ancient Vietnamese symbol of peace, disappeared during the war's ecological holocaust, along with the white-winged duck, the giant ibis, and the milky stork. After the war, international and Vietnamese environmentalists worked with local families to build dikes to keep fresh rainwater in the fields, and to replant the cajeput and mangrove trees as well as restore wetland grasses around the fields to keep the water clean. The nation set up the Tram Chim National Park, which was patrolled by armed guards and local families to ensure its success. It took a long time and a lot of effort, but the land and the people started to heal, and the rice harvests improved. Decades later, the people started spotting the cranes again. They had returned.

This is a story of hope and the power of resilience and restoration after a war, showing that the quality of life for both the people and nature are inherently linked. As terrible as this war was, not all countries are even as fortunate as Vietnam in their rebuilding. Pope John Paul II pronounced that "wars are among the worst causes of environmental damage" (PONTIFICAL ACADEMY STUDY WEEK ON RESOURCES AND POPULATION, NOVEMBER 22, 1991).

Tabulating the Ecological Damage

During the first Gulf War, irradiated weaponry was used—from 290 to 800 tons of depleted uranium projectiles. The cost of trying to remove the contamination has been 4 to 5 billion dollars per 200 hectares. Over 700 oil rigs were set afire, sending tons and tons of carbon and pollution into the atmosphere and spilling 60 million barrels of oil. Kuwait's soil and its main groundwater aquifer remained contaminated years later. The Kuwait Bay received the fallout from the burning oil wells, leaving marine habitats polluted by a surface oil sludge. Many sewage treatment plants were damaged, with just one of them spewing 50,000 cubic meters of raw sewage a day into the ocean until the plant could be rebuilt. This is to say nothing of the landmines and the long-term cancers and other diseases that result from the chemical and irradiated pollutants and weaponry. These are just a smattering of the long-term ecological effects from one recent, relatively small-scale war.

What of civil wars and ethnic conflicts that rage for decades, doing their damage? Environmental catastrophes, deforestation, and drought are often the immediate consequences of contemporary warfare. Pope John Paul II reminded the world also that "wars and internal conflicts are the primary cause of food shortages," and famine (PASTORAL VISIT TO ASSISI, OCTOBER 4, 2000).

Sadly, there are also the horrific immediate human costs of war—not just the deaths of soldiers and civilians but the creation of entire nations of refugees who have fled the only homes they have ever known. Globally, there are an estimated 15.2 million refugees who have left their countries because of war, with another 26.4 who are refugees in new areas within their own countries.

Refugee camps, besides being beset by safety problems (with rape and loot-ing rampant), present ecological problems as well, as the refugees face scarcity of food, water, and shelter. The concentrated sewage and waste create additional ecological damage that results in public health disasters and river pollution.

Having lived through a war in Poland, John Paul II knew better than most some of the real costs that governments, politicians, generals and the public don't consider in their war rooms. He'd also witnessed the results of wars in various continents and saw how widespread ecological damage made it so much harder to rebuild lives:

> Today, any form of war on a global scale would lead to incalculable ecological damage. But even *local or regional wars,* however limited, not only destroy human life and social structures, but *also damage the land, ruining crops and vegetation as well as poisoning the soil and water.* The survivors of war are forced to begin a new life in very difficult envi-ronmental conditions, which in turn create situations of extreme social unrest, with further negative consequences for the environment.

> MESSAGE FOR WORLD DAY OF PEACE, "PEACE WITH GOD THE CREATOR, PEACE WITH ALL OF CREATION," JANUARY 1, 1990

Children as Casualties

The pope's heart particularly went out to the young victims of war. Their bodies absorb the chemical toxins and their growth is stunted by the malnu-trition; they are orphaned or made into refugees; they are kidnapped to be child soldiers, some forced to commit atrocities against their own families, many later drowning their sorrow and guilt in alcohol and drugs. They are trained into a culture of rape, robbery, fear, and death. In impoverished coun-tries with ongoing war, young men often take up freelance soldiering and robbery as a kind of profession, since this is all they've ever learned.

After a group of gunmen killed 186 schoolchildren in Beslan, Russia, Pope John Paul II reminded the world of all the children harmed by violence:

> At this moment our gaze broadens to take in *all innocent children* in every corner of the earth who are victims of the violence of adults.

Children *forced to use weapons and taught to hate and kill;* children *induced to beg in the streets,* exploited for easy earnings; children *ill-treated and humiliated* by arrogant, abusive grown-ups; children *left to themselves,* deprived of the warmth of a family and prospects of a future; children who *die of hunger,* children *killed in the many wars in various regions of the world.*

It is a loud cry of pain from children whose dignity is offended. It cannot, it must not leave anyone indifferent.

GENERAL AUDIENCE, SEPTEMBER 8, 2004

How often John Paul II said to the people and leaders of the world, stop being indifferent! Stop shutting your eyes and closing your ears to the cries of the poor, especially those of the children living among violence in destroyed environments. In his peace message "Do Not Be Overcome By Evil, But Overcome Evil with Good," he quoted a line of his own he often used:

To attain the good of peace there must be a clear and conscious acknowledgment that violence is an unacceptable evil and that it never solves problems. "Violence is a lie, for it goes against the truth of our faith, the truth of our humanity. Violence destroys what it claims to defend: the dignity, the life, the freedom of human beings."

WORLD DAY OF PEACE 2005 MESSAGE, "DO NOT BE OVERCOME BY EVIL, BUT OVERCOME EVIL WITH GOOD," JANUARY 1, 2005

So the pope said, *tell the truth, the whole truth, about the costs of war on every level, from the human to the wildlife, to the systems of land, water, and air, and to the future generations.*

The Ecological Costs of Long-Lasting Weapons of Destruction

When the general public thinks about war, they tend to think that after the conflict ends, so does the destruction. That is hardly the case with modern weapons.

In 1999, Pope John Paul II spoke out against the vast number of land-mines "which have maimed or killed hundreds of thousands of innocent people, while *despoiling fertile land*," calling for all nations to join in the international action against them (*ECCLESIA IN ASIA*, 1999). Two years earlier, an International Treaty for a Ban on Landmines had been signed in Ottawa by 122 nations, helped along by the adamant support of the pope and the Holy See. Currently it has 161 signers, including all the countries in Europe and the Americas *except* the United States and Cuba. (John Paul II praised Canada for its leadership on the ban.) Many significant nations in Asia and the Middle East also have still not signed (as of October 2013), including India, Saudi Arabia, Israel, Russia, Vietnam, North Korea, and South Korea.

As is well known, some of the most devastating human and environ-mental violence in history was the result of dropping atomic bombs on Hiroshima and Nagasaki. Pope John Paul II reminded the United Nations of this at its 50th anniversary celebration in 1995, and he thanked God that five decades had passed without a recurrence of nuclear weapon use. He attrib-uted this in part to the dialogue, mediation, and sanctions organized through the UN, as well as to the effectiveness of its programs in so many areas of human development and ecological concern:

> …the second half of the Twentieth Century has seen the unprec-edented phenomenon of a humanity uncertain about the very likelihood of a future, given the threat of nuclear war. That danger, mercifully, appears to have receded—and everything that might make it return needs to be rejected firmly and universally; all the same, fear for the future and of the future remains.
>
> ADDRESS TO GENERAL ASSEMBLY OF THE UNITED NATIONS, NEW YORK, OCTOBER 5, 1995

But nuclear and irradiated weaponry are not the only game changers when it comes to modern warfare, and John Paul warned that military forces were playing with the very fundamentals of life as they developed new forms of waging war:

There is another dangerous menace that threatens us, namely *war.*

Unfortunately, modern science already has the capacity to *change the environment for hostile purposes.* Alterations of this kind over the long term could have unforeseeable and still more serious consequences. Despite the international agreements that prohibit chemical, bacteriological and biological warfare, the fact is that laboratory research continues to develop new offensive weapons capable of altering the balance of nature.

WORLD DAY OF PEACE MESSAGE "PEACE WITH GOD THE
CREATOR, PEACE WITH ALL CREATION," JANUARY 1, 1990

The world was reminded of this again in 2013 when Syria launched a chemical weapons attack against its own civilians to squelch the uprising. In response, Pope Francis issued an unequivocal statement against chemical weapons:

With utmost firmness I condemn the use of chemical weapons: I tell you that those terrible images from recent days are burned into my mind and heart. There is a judgment of God and of history upon our actions, which are inescapable! Never has the use of violence brought peace in its wake.

ANGELUS, SEPTEMBER 1, 2013

Responding to world pressure, Syria signed and ratified the Chemical Weapons Convention on October 14, 2013. The international community now has to exert similar pressure on those nations still outstanding—as of fall 2013, Israel, Egypt, North Korea, Angola, South Sudan, and Burma have not yet ratified the treaty.

War's Effects on the Domestic Budgets

War is the most expensive enterprise a country can undertake, rearranging national priorities, goals, and budgets, and well as those of individuals and families. As the costs mount, funds are diverted from other areas such as conservation, job stimulus, health care, and education. This disturbed Pope John Paul II, as he expressed to the bishops of Asia in 1999:

> Especially troubling in Asia is the continual race to acquire weapons of mass destruction, an immoral and wasteful expenditure in national budgets, which in some cases cannot even satisfy people's basic needs....
>
> The Synod called for a stop to the manufacture, sale and use of nuclear, chemical and biological arms and urged those who have set landmines to assist in the work of rehabilitation and restoration.
>
> POST-SYNODAL APOSTOLIC EXHORTATION, *ECCLESIA IN ASIA*, 1999

The pope preached this same message to all countries, including those in America. He called on the United States to disarm and rearrange its budget priorities toward those programs and systems that are restorative and can make a positive difference in people's lives:

> One factor seriously paralyzing the progress of many nations in America is the arms race. The particular Churches in America must raise a prophetic voice to condemn the arms race and the scandalous arms trade, which consumes huge sums of money which should instead be used to combat poverty and promote development. On the other hand, the stockpiling of weapons is a cause of instability and a threat to peace.
>
> POST-SYNODAL APOSTOLIC EXHORTATION, *ECCLESIA IN AMERICA*, 1999

The arms trade, tests, and stockpiling (of the "military-industrial complex" as President Eisenhower termed it) results in more pollution and ecological damage at army bases at home and abroad. Investigative reports claim that the United States military has been one of the planet's greatest polluters, though it is presently working on reform. In 2010, the online column "Earthtalk," from *E/The Environmental Magazine,* reported the findings of a nonprofit dedicated to carefully covering the stories that corporate news media neglect:

According to the nonprofit watchdog group, Project Censored, American forces generate some 750,000 tons of toxic waste annually—more than the five largest U.S. chemical companies combined. Although this pollution occurs globally on U.S. bases in dozens of countries, there are tens of thousands of toxic "hot spots" on some 8,500 military properties right here on American soil.... The nonprofit Military Toxics Project is working with the U.S. government to identify problem sites and educate neighbors about the risks....

Meanwhile, the U.S. military manages 25 million acres of land that provides habitat for some 300 threatened or endangered species. The military has harmed endangered animal populations by bomb tests (and been sued for it), reports Project Censored, and military testing of low-frequency underwater sonar technology has been implicated in the stranding deaths of whales worldwide. Despite being linked to such problems, the U.S. Department of Defense (DoD) has repeatedly sought exemptions from Congress for compliance with federal laws including the Migratory Bird Treaties Act, the Wildlife Act, the Endangered Species Act, the Clean Air Act and the National Environmental Policy Act.

NOVEMBER 1, 2010, POLITCALAFFAIRS.NET

This accounting does not including the greenhouse gas emissions of the many military tanks, trucks, jeeps, hummers, planes, ships, bases, and munitions factories.

On the hopeful side, "Earthtalk" also reported on what the U.S. military is doing to clean up its act, especially regarding its contribution to greenhouse gas pollution:

A recent Obama administration directive calls for the DoD to draw 20% of its power from renewable sources by 2020. Nikihl Sonnad of the GreenFuelSpot website reports that the Army and Air Force are planning to include solar arrays on several bases in sunny western states. The Air Force is also building the nation's largest biomass energy plants in Florida and Georgia, and the Navy is building three large geothermal energy plants and funding research into extracting energy from ocean waves.

NOVEMBER 1, 2010, POLITCALAFFAIRS.NET

Ecological Causes of War

Pope John Paul II taught that war, whether considered "just" or not, exhibits a failure of anticipation, thinking, life values, and conflict resolution, which only leads to more violence and less respect for all life, especially nature. War creates problems on every level—social, physical, and ecological. Yet the reverse is also true. Problems in all these areas, *especially ecological destruction, can cause wars.*

The pope knew well how ecological and natural resource destruction can send societies rushing like lemmings over the cliff to war:

> In our day, there is a growing awareness that world peace is threatened not only by the arms race, regional conflicts, and continued injustices among peoples and nations, but also by a lack of *due respect for nature, by the plundering of natural resources and by a progressive decline in the quality of life.*

WORLD DAY OF PEACE 1990 MESSAGE "PEACE WITH GOD THE CREATOR, PEACE WITH ALL CREATION," JANUARY 1, 1990

In *Collapse: How Societies Choose to Fail or Succeed,* scientist Jared Diamond traces how various societies (including the Anasazi, Mayans, Easter Islanders, and the Greenland Norse) fell apart, disappeared, or ended up in destabilizing wars because they failed to pay attention to the limits of their land, water, soil, or wildlife or because they refused to change entrenched habits and practices in the face of problems. In other words, they refused to change even in the midst of mounting evidence that what they were doing and thinking was wrong, and wouldn't work in the long run. When reasoning was used to look for solutions, it was poorly processed. The leaders considered the wrong information important or applied "disastrous values" as Diamond termed them. In these times of ecological crisis, the histories he recounts serve as chilling cautionary tales.

John Paul II recognized the consequences of the disastrous values of short-term use, greed, disposability and waste, and sought to steer the world instead to apply values of life and long-term vision, recognizing the inherent connections between ecological destruction and war, and vice versa.

Seeking Peaceful Answers to Conflicts

Over and over, from every angle, Pope John Paul II drummed out the message: war is *not* the answer.

> She [the Church] is convinced that war creates more problems than it ever solves, that dialog is the only just and noble path to agreement and reconciliation, and that the patient and wise art of peacemaking is especially blessed by God.
>
> POST-SYNODAL APOSTOLIC EXHORTATION, *ECCLESIA IN ASIA*, 1999

His words, though, were never a negative reflection on those who served nations as soldiers or the world as trained peacekeepers. He recognized that violent bullies need to be stopped, and that this often requires bystander countries and outside mediators to intervene using both non-violent tools as well as guardian forces to serve as defenders. When the pope spoke to soldiers and peacekeeping forces, he blessed and thanked them:

> The work and sacrifice of all of you help to ensure the peace and security of individuals and societies. I pray that you yourselves will always be kept safe as you fulfill your professional duties, and that the divine gifts of wisdom and strength will ever accompany you in the service of your own countries and of your fellow men and women.
>
> ANGELUS, NOVEMBER 19, 2000

The brave service of soldiers and international peacekeepers puts even more moral responsibility on national and world leaders never to put them in harm's way unnecessarily, and never to use them for preemptive strikes or any efforts that are not for the defense of the defenseless, and never to protect commerce, sources of profit, or "national interests"—only to protect life.

Towards a Grammar of Moral Law and Vocabulary of Values

To reduce the rise of war from "disastrous values," Pope John Paul II proposed to the General Assembly at its 50th anniversary that the United Nations needs to work together to craft a *universal grammar of moral law*. Such a mechanism would articulate shared principles of ethics, philosophy, and religion that could facilitate deeper dialogue on every issue. In discussions of key issues, representatives could evoke "moral principles inscribed on human consciences" as the pope described them. With this shared vocabulary of values outlining accepted "rules for individual and group action and the reciprocal relationships of persons and nations in accord with justice and solidarity," people *could count on being understood and respected* beyond borders of culture, religion, language, or nation. Such a grammar of moral law would be used to work towards peace and the common good, encourage respect among all people, and protect all of life and the systems upon which it depends.

The Universal Declaration of Human Rights was the first element of this moral structure, proving that it can be constructed. As John Paul II explained, these agreed–upon human rights:

> . . . *remind us that we do not live in an irrational or meaningless world*. On the contrary, there is a *moral logic* which is built into human life and which makes possible dialogue between individuals and peoples. . . .
>
> If we want a *century of violent coercion* to be succeeded by a *century of persuasion*, we must find a way to discuss the human future intelligibly. The universal moral law written on the human heart is precisely that kind of "grammar" which is needed if the world is to engage this discussion of its future.

> ADDRESS TO THE 50TH GENERAL ASSEMBLY OF THE
> UNITED NATIONS, NEW YORK, OCTOBER 5, 1995

After his death, Pope Benedict XVI passionately pursued this same UN proposal as a path to peace, quoting his friend and mentor, John Paul II.

Honoring the Diversity in God's People

To accomplish this international consensus on moral values, different perspectives—of nations, cultures, religions, race, gender and sexual orientation, age, appearance, ability, and resources—must be acknowledged and respected. It is one thing to glory in the earth's biodiversity, it is another to delight in its diversity of its people. To honor all God's creation, one must.

For Pope John Paul II, handling diversity was just another chance for gratitude and learning—for with it, one could absorb a greater understanding of the Creator and of humanity's full meaning and potential.

> For different cultures are but different ways of facing the question of the meaning of personal existence. . . . every culture is an effort to ponder the mystery of the world and in particular, of the human person: it is a way of giving expression to the transcendent dimension of human life. *The heart of every culture is its approach to the greatest of all mysteries: the mystery of God. . . .*
>
> To cut oneself off from the reality of difference—or, worse, to attempt to stamp out that difference—is to cut oneself off from the possibility of listening to the depths of the mystery of human life. . . . every culture has something to teach us about one or other dimension of that complex truth.
>
> *Thus the "difference" that some find so threatening can, through respectful dialog, become the source of a deeper understanding of the mystery of human existence.*
>
> ADDRESS TO THE 50TH GENERAL ASSEMBLY OF THE
> UNITED NATIONS, NEW YORK, OCTOBER 5, 1995

Preventing War and Global Terrorism through Religious Respect and Cooperation

Throughout history, conflicts between religions have been one of the key factors in sparking wars. In October 1986, in a totally unprecedented act, Pope John Paul II called a gathering of religious leaders from around the world to pray together for peace. Not surprisingly, he picked St. Francis's hometown, Assisi, as the site. Leaders from over sixty different religions gathered: Catholics, all other branches of Christianity, Jews, Buddhists, Shintoists, Muslims, Zoroastrians, Hindus, Unitarians, members of traditional African and Native American religions, and many others. Together, under the roof of the Basilica of St. Francis, they prayed, side by side, in their own faith traditions, for world peace. At the end, John Paul II addressed them all and thanked them for coming:

> For the first time in history, we have come together from everywhere, Christian Churches and Ecclesial Communities, and World Religions, in this sacred place dedicated to St. Francis, to witness before the world, each according to his own conviction, about the transcendent quality of peace.
>
> The form and content of our prayers are very different, as we have seen, and there can be no question of reducing them to a kind of common denominator. *Yes, in this very difference we have perhaps discovered anew that, regarding the problem of peace and its relation to religious commitment, there is something which binds us together.* . . .
>
> . . . the conviction that peace goes much beyond human efforts, particularly in the present plight of the world, and therefore that its source and realization is to be sought in that Reality beyond all of us. . . .

> OCTOBER 26, 1986

Pope John Paul II made it clear that to be listening and respectful and sharing in prayer does *not* mean one has to be silent about one's own beliefs or be rootless, accepting all dogmas or beliefs as one's own. Instead, it means to accept others as they are and only offer one's own beliefs as a gift of sharing, ready to accept others in the same fashion, with no imposition, coercion, condescension, or disrespect. He ended his Assisi conference:

> That is why each of us prays for peace. Even if we think, as we do, that the relation between that Reality and the gift of peace is a different one, according to our respective religious convictions, we all affirm that such a relations exists. This what we express by praying for it.

I humbly repeat here my own conviction: peace bears the name
of Jesus Christ.

ASSISI, OCTOBER 26, 1986

Coping With and Appreciating Difference

Pope John Paul II felt absolutely convinced that without such interfaith
respect and exchanges, there would be no possibility of dealing with questions of global scope, like world peace, climate change, and poverty. The
very act of the pope praying with non-Christians, however, infuriated many
Christians and Roman Catholics around the world, even those close to him,
such as Cardinal Joseph Ratzinger. Pope John Paul II cautioned everyone to
remember a key truth:

> God is not at the beck and call of one individual or one people, and
> no human venture can claim to monopolize him. The children of
> Abraham know that God cannot be commandeered by anyone: God is
> to be received.

ADDRESS TO THE DIPLOMATIC CORPS, JANUARY 10, 2002

In all interfaith exchanges, John Paul II established guidelines that fit
the Creator's model of nondiscriminatory love for all people and species:

> Interreligious relations are best developed in a context of openness to
> other believers, a willingness to listen, and the desire to respect and
> understand others in their differences. For all this, love of others is
> indispensable. This should result in collaboration, harmony, and mutual
> enrichment.

POST-SYNODAL APOSTOLIC EXHORTATION,
ECCLESIA IN ASIA, QUOTING SYNOD FATHERS, 1999

Pope Francis put it even more bluntly in his interview with the Italian
journalist, and atheist, Eugenio Scalfari. The pope said that sharing one's faith
is not about proselytizing to convert someone but visibly passing on Christ's
love through open exchange: "Proselytism is solemn nonsense, it makes no

sense. We need to get to know each other, listen to each other and improve our knowledge of the world around us." He elaborated:

> It is love of others, as our Lord preached. It is not proselytizing, it is love. Love for one's neighbor, that leavening that serves the common good.... I have already said that our goal is not to proselytize but to listen to needs, desires and disappointments, despair, hope.

LA REPUBBLICA, OCTOBER 1, 2013

The Role of Religions to Nurture Peace

Fifteen years after the first Assisi interfaith gathering came the devastating events of September 11, 2001. Pope John Paul II was particularly shocked and disheartened that this and so many other terrorist acts were said to be done "in the name of God." He turned once more to the model of St. Francis, and to the city the saint blessed, to try to build peace. This time, Cardinal Ratzinger was more approving. (Later, as Pope Benedict XVI, he modeled his predecessor by convening the next interfaith gathering for peace at Assisi in October 2011, to mark the 25th anniversary of the first one and the upcoming 10th anniversary of 9/11).

When the religious representatives arrived in Assisi in January 2002, John Paul II thanked them for coming and clarified their mission:

> It is the *duty of religions,* and of their leaders above all, to foster in the people of our time *a renewed sense of the urgency of building peace....*
>
> Tragic conflicts often result from *an unjustified association of religion with nationalistic, political and economic interests*, or concerns of other kinds. Once again, gathered here together, we declare that *whoever uses religion to foment violence contradicts religion's deepest and truest inspiration.*
>
> It is essential, therefore, that *religious people and communities should in the clearest and most radical way repudiate violence, all violence, starting with the violence that seeks to clothe itself in religion,* appealing even to the most holy name of God in order to offend man.... There is no religious goal that can possibly justify the use of violence by man against man.

ADDRESS TO THE REPRESENTATIVES OF THE WORLD RELIGIONS,
DAY OF PRAYER FOR PEACE IN THE WORLD, ASSISI, JANUARY 24, 2002

In good faith and with a view to the common good, the religious leaders worked out ten interfaith principles for building peace between religions, cultures, and nations. The "Assisi Decalogue for Peace" is "the inspired reflections of these men and women, representatives of different religious confessions, in their sincere desire to work for peace, and their common quest for the true progress of the whole human family":

ASSISI DECALOGUE FOR PEACE

We commit ourselves to:...

1. Proclaiming our firm conviction *that violence and terrorism are incompatible with the authentic spirit of religion*, and, as we condemn every recourse to violence and war in the name of God or of religion, we commit ourselves to doing everything possible to eliminate the root causes of terrorism.
2. *Educating people to mutual respect and esteem*, in order to help bring about a peaceful and fraternal coexistence between people of different ethnic groups, cultures and religions.
3. *Fostering the culture of dialog...*
4. *Defending the right of everyone to live a decent life in accordance with their own cultural identity, and to form freely a family of his own.*
5. *Frank and patient dialog, refusing to consider our differences as an insurmountable barrier, but recognizing instead that to encounter the diversity of others can become an opportunity for greater reciprocal understanding.*
6. *Forgiving one another for past and present errors and prejudices, and to supporting one another in a common effort both to overcome selfishness and arrogance, hatred and violence*, and to learn from the past that peace without justice is no true peace.
7. *Taking the side of the poor and the helpless...*
8. Taking up the cry of those who refuse to be resigned to violence and evil, and we are desire to make every effort possible to *offer the men and women of our time real hope for justice and peace.*
9. *Encouraging all efforts to promote friendship between peoples,* for we are convinced that, in the absence of solidarity and understanding

between peoples, technological progress exposes the world to a
growing risk of destruction and death.

10. *Urging leaders of nations to make every effort to create and consolidate, on the national and international levels, a world of solidarity and peace based on justice.*

LETTER OF JOHN PAUL II TO ALL THE HEADS OF STATE AND GOVERNMENT
OF THE WORLD, ASSISI, JANUARY 2002, DATED FEBRUARY 4, 2002

Moral Dialogue Plus Disarmament

If a gun is aimed at you, and you're aiming a gun at someone else, it's difficult to build a sense of mutual respect or mediation, or live out the "Decalogue for Peace." Respectful moral dialogue must be built upon mutual disarmament. Pope John Paul II pled for disarmament in all his apostolic letters, such as this one to bishops and national leaders in Asia:

> It is the responsibility of all, *especially of those who govern nations, to work more energetically for disarmament....*
>
> Above all the Synod Fathers prayed to God... to put sentiments of peace in the hearts of those tempted to follow the ways of violence so that the Biblical vision will become a reality: "they shall beat their swords into ploughshares, and their spears into pruning hooks; nation shall not lift up sword against nation, neither shall they learn war any more" (*Is* 2:4).

POST-SYNODAL, APOSTOLIC EXHORTATION, *ECCLESIA IN ASIA*, 1999

He came down hard on nations in the Americas, particularly the United States, for gathering up more and more weapons rather than disarming, and for often leaning in favor of waging distant wars to solve problems. In his 1999 *Ecclesia in America*, he stated bluntly:

> For this reason the Church remains vigilant in situations where these is a risk of armed conflict, even between sister nations. As a sign and instrument of reconciliation and peace, she must seek "by every means possible, including mediation and arbitration, to act in favor of peace and fraternity between peoples."

With this in mind, John Paul II worked actively to prevent the preemptive Gulf Wars, fearing the human and ecological disasters and suffering that would result.

The Gulf Wars: Case Studies of Principles

In 1991, as President George H. W. Bush was considering invading Kuwait, Pope John Paul II observed:

> The anxiety and sadness that unfortunately have been voiced more than once over the war in the Gulf region continue to be increased by the prolonged fighting: catastrophic ecological risks are now added to them.
>
> ANGELUS, JANUARY 27, 1991

In the end, the first Gulf War clearly did not fix all the problems President Bush, Congress, and the U.S. allies had sought to address.

After 9/11 and the interfaith prayer gathering at Assisi, Pope John Paul II, though increasingly ravaged by Parkinson's and illness, committed himself to doing all he could to intervene and try to stop a second Gulf War and potential escalation. At the same time, he was deeply saddened and distressed by terrorism and the ongoing conflicts in the nations of Africa and other areas—and all the ecological crises that were brought along with them. So in 2003, weak in body but bold in faith, he empowered his international diplomatic corps to push for nonviolent progress on conflicts and a transformation in attitudes and habits toward the environment:

> I have been personally struck by the feeling of fear which often dwells in the hearts of our contemporaries. An insidious terrorism capable of striking at any time and anywhere...the conflicts preventing numerous African countries from focusing on their development; the diseases spreading contagion and death; the grave problem of famine, especially in Africa; *the irresponsible behavior contributing to the depletion of the planet's resources:* all these are so many plagues *threatening the survival of humanity,* the peace of individuals and the security of societies.

Yet everything can change. It depends on each of us. Everyone can develop within himself his potential for faith, for honesty, for respect of others, and for commitment to the service of others.

It also depends, quite obviously, *on political leaders*, who are called to serve the common good. You will not be surprised if before an assembly of diplomats I state...certain requirements that I believe must be met *if entire peoples, perhaps even humanity itself, are not to sink into the abyss.*

First, a "YES TO LIFE"!....

Next, RESPECT FOR LAW........

Finally, the DUTY OF SOLIDARITY....

This is why *choices need to be made so that humanity can still have a future.* Therefore, the peoples of the earth and their leaders must sometimes have the courage to say "No".

"NO TO DEATH"!...

"NO TO SELFISHNESS"! In other words, to all that impels people to protect themselves inside the cocoon of a privileged social class or a cultural comfort that excludes others. *The life-style of the prosperous, their patterns of consumption, must be reviewed in the light of their repercussions on other countries.*

Let us mention for *example the problem of water resources,* which the United Nations Organization has asked us all to consider during this year 2003. *Selfishness is also the indifference of prosperous nations towards nations left out in the cold.* All peoples are entitled to receive a fair share of the goods of this world and of the know-how of the more advanced countries....

"NO TO WAR"! War is not always inevitable. *It is always a defeat for humanity.* International law, honest dialogue, solidarity between States, the noble exercise of diplomacy: these are methods worthy of individuals and nations in resolving their differences.

I say this as I think of those who still place their trust in nuclear weapons and of the all-too-numerous conflicts that continue to hold hostage our brothers and sisters in humanity....

As the Charter of the United Nations organization and international law itself remind us, war cannot be decided upon, even when it is a matter of ensuring the common good, *except as the very last option and in accordance with very strict conditions, without ignoring the consequences for the civilian population both during and after the military operations.*

It is therefore possible to change the course of events, once good will, trust in others, fidelity to commitments and cooperation between responsible partners are allowed to prevail.

ADDRESS TO THE DIPLOMATIC CORPS, JANUARY 13, 2003

The Showdown in the Gulf

His outlining of a new path did little to redirect the clash of wills ensuing between U.S. President George W. Bush and Saddam Hussein. As things grew more tense and the United States had started to deploy soldiers to the Persian Gulf, the pope sent an envoy with a personal letter to each leader to try to find a way to make peace. John Paul II cautioned President Bush not to launch a preemptive strike and invade Iraq:

> I wish now to restate my firm belief that war is not likely to bring an adequate solution to international problems and that, even though an unjust situation might be momentarily met, the consequences that would possibly derive from war would be devastating and tragic. We cannot pretend that the use of arms, and especially of today's highly sophisticated weaponry, would not give rise, in addition to suffering and destruction, to new and perhaps worse injustices.

DELIVERED BY SPECIAL ENVOY CARDINAL ROGER ETCHEGARAY TO PRESIDENT HUSSEIN, FEBRUARY 15, 2003; CARDINAL PIO LAGHI TO PRESIDENT GEORGE W. BUSH MARCH 5, 2003.

In February, the pope met with Prime Ministers Tony Blair and Silvio Berlusconi, and it was rumored that John Paul II lost his temper during this meeting, an unusual occurrence for him. One paper reported he "raised his voice, pointed an accusing finger...and even banged his fist on the table." On March 16, 2003, the pope addressed the world from St. Peter's Square:

> Certainly the political leaders in Baghdad have the urgent duty to collaborate fully with the international community to eliminate

all motives for military intervention. But I also want to remind the member states of the United Nations, and in particular those in the Security Council, that the use of force represents the last resort, after all other peaceful solutions have been exhausted, according to the principles of the UN Charter itself.

Two days later, as the international game of chicken continued, the pope issued the following press statement: "Whoever decides that all peaceful methods of international law have been exhausted assumes a grave responsibility before God, his conscience, and history."

When the pope heard of the invasion of Iraq, Cardinal Roberto Tucci, who was a close friend of the pope's and his confidant and spokesperson at times, called the U.S. attack "a defeat for reason and the gospel. The war is beyond all legality and international legitimacy." Tucci's words became ascribed to the pope himself. Such were the lengths Pope John Paul II went to to try to stop the two Gulf Wars.

Pope Francis Follows His Example in Working for Peace

Pope Francis, too, has followed the path of his namesake, and of John Paul II. On September 1, 2013, when it seemed that the United States was ready to begin strikes on Syria to punish it for using chemical weapons, Francis called for an international day of fasting and prayer:

> I have decided to proclaim for the whole Church September 7, the vigil of the birth of Mary, Queen of Peace, a day of fasting and prayer for peace in Syria, the Middle East, and throughout the world, And I also invite each person, including our fellow Christians, followers of other religions and all men of good will, to participate, in whatever way they can, in this initiative....
>
> We will gather together in prayer, in a spirit of penitence, to ask from God this great gift for the beloved Syrian nation and for all the situations of conflict and violence in the world.
>
> ANGELUS, SEPTEMBER 1, 2013

He exhorted all parties to negotiate and find a way to stop the use of chemical weapons and the Syrian war without drone strikes.

The following week, Secretary of State John Kerry made a chance comment about Syria disposing of its chemical weapons to avoid a strike, and the suggestion was pounced upon as a nonviolent solution by his Russian counterpart, Foreign Minister Sergei Lavrov. This encounter launched negotiations to encourage Syria to sign the Chemical Weapons Convention and invite in international inspectors and advisors to eliminate its stockpile.

This was an incredible triumph of dialogue over violence. Intense world scrutiny and new tools of international pressure are still needed in Syria, other Mideast countries, and many other places to stop the conventional attacks on civilians and human rights violations. But a new model of solidarity and dialogue for peace was established.

Make Audacious Acts to Build Peace

For real peacemaking at all levels of our lives, Pope John Paul II believed we have to, like Jesus, risk "audacious" acts to disarm others with kindness, respect, and gestures of listening and good will. We will have to build solidarity, strength of character, and courage, and work to be our higher selves individually, religiously, communally and internationally, seeking the common good and a livable planet for the present and future. In his initial 1979 World Day of Peace message, "To Reach Peace, Teach Speech," he laid out the obligations of national and international leaders to God, their countries, and the world:

> Leaders of the peoples, *learn to love peace by finding in the great pages of your national histories... your predecessors whose glory lay in giving growth to the fruits of peace.* "Blessed are the peacemakers."...
>
> Statesmen
>
> Open up new doors to peace. Do everything in your power to make the way of dialog prevail over that of force....Be convinced that honor and effectiveness in negotiating...are not measured by the degree of inflexibility in defending one's interests, but by the participants' capacity for respect, truth, benevolence and brotherhood or, let us say, by their humanity.

Make gestures of peace, even audacious ones, to break free from vicious circles and from the deadweight of passions inherited from history. Then patiently weave the political, economic, and cultural fabric of peace....

By making resolute gestures of peace you will release the true aspirations of the peoples and will find in them powerful allies in working for the peaceful development of all.

WORLD DAY OF PEACE 1979 MESSAGE, "TO REACH PEACE, TEACH PEACE," JANUARY 1, 1979

PART IV

ECOLOGICAL CONVERSION
AND RESTORATION

It is not too late,
God's world has incredible healing powers. Within a simple generation,
we could steer the earth toward our children's future.
Let that generation start now,
with God's help and blessing.

POPE JOHN PAUL II WITH ECUMENICAL PATRIARCH BARTHOLOMEW I,
COMMON DECLARATION ON THE ENVIRONMENT, ROME-VENICE, JUNE 10, 2002

On the basis of the covenant with the Creator...
each one is invited to a deep personal conversion
in his or her relationship with others and with nature.

ADDRESS TO THE PONTIFICAL ACADEMY OF SCIENCES, STUDY WEEK ON
SCIENCE FOR SURVIVAL AND SUSTAINABLE DEVELOPMENT, MARCH 12, 1999

Conversion in Our Relationships with Our Creator and All of Creation

In 2002, Pope John Paul II, with Ecumenical Patriarch Bartholomew I, wrote in a joint message to the world:

> We must frankly admit that humankind is entitled to something better than what we see around us. We and much more our children and future generations are entitled to a better world, a world free from degradation, violence and bloodshed, a world of generosity and love.

COMMON DECLARATION ON ENVIRONMENTAL ETHICS, POPE JOHN PAUL II AND
THE ECUMENICAL PATRIARCH BARTHOLOMEW I, ROME-VENICE, JUNE 10, 2002

So how do we meet the twelve related crises?

With humble repentance, pure and simple, said Pope John Paul II. We cannot change if we cannot own up to the damage done and the suffering caused by our choices. With Ecumenical Patriarch Bartholomew, he stated:

> What is required *is an act of repentance* on our part and a renewed attempt to view ourselves, one another, and the world around us within the perspective of the divine design for creation.
>
> The problem is not simply economic and technological; it is moral and spiritual. A solution at the economic and technological level can be found only if we undergo, in the most radical way, an inner change of heart, which can lead to a change in lifestyle and

unsustainable patterns of consumption and production. A genuine *conversion* in Christ will enable us to change the way we think and act.

First, we must regain our humility and recognize the limits of our powers, and most importantly, the limits of our knowledge and judgment. We have been making decisions, taking actions, and assigning values that are leading us away from the world as it should be, away from the design of God for creation, away from all that is essential for a healthy planet and a healthy commonwealth of people.

COMMON DECLARATION ON ENVIRONMENTAL ETHICS, POPE JOHN PAUL II AND
THE ECUMENICAL PATRIARCH BARTHOLOMEW I, ROME-VENICE, JUNE 10, 2002

Faith-Filled Ecological Repentance and *Metanoia*

What does this ecological repentance mean in everyday terms? The word "conversion" is the English translation of the Greek word *metanoia*. Pope John Paul II explained that *metanoia:*

> literally means to allow the spirit to be overturned in order to make it turn toward God. These are also the two fundamental elements which emerge from the parable of the son who was lost and found: his "coming to himself" and his decision to return to his father....
>
> It means the inmost change of heart under the influence of the word of God and in the perspective of the kingdom.

POST-SYNODAL APOSTOLIC EXHORTATION, "RECONCILIATION
AND PENANCE," DECEMBER 2, 1984

The ecological conversion needed must be one of *metanoia*, a turning inside out of our values for a "coming home" to our humble place in the universe—to see "with new eyes" the importance of nature as part of who we are and reorient ourselves and how we act in relationship to all of God's creations. John Paul used this kind of image when speaking with youth from all over the world at Viterbo, Italy:

> Another aspect of Christ's newness that you are to witness before the world: *a new relationship with the environment surrounding you.* Above

all in this past century, people have made a use of the earthly realities which....has been irresponsible. *It is necessary to learn to look at nature with new eyes....*

...The believer's attitude does not make a foolish use of resources or use arbitrary violence towards the animal kingdom to which all people belong in bodily dimension....

ADDRESS TO YOUTH IN VITERBO, ITALY, MAY 27, 1984

How Do We Achieve This Reorientation?

When Jesus walked the earth, two blind men came to him and asked for mercy. He looked at them and said, "What do you want me to do for you?" They said in reply, "Lord, let our eyes be opened." The Gospel goes on to say that, "Moved with compassion, Jesus touched their eyes. Immediately they regained their sight and followed him" (MATTHEW 21:31-34). The men could suddenly see not only what was right in front of them but the whole context of where they lived and many things at a distance, beyond touch. Their worlds were vastly expanded beyond themselves, giving them new understandings, meanings, and range. They could also see people's expressions and body language, and suddenly understood other people's feelings and needs in new ways.

This is what John Paul II was asking the young people, and all people, to seek—that our eyes, minds, and hearts be opened to how our present ecological crises are intertwined with our spiritual, social, and economic actions; that we seek to become educated about the issues and willing to respond on our own and in religious and civic groups and as nations; and that we see the opportunities for growth at all levels in doing so.

Love Is the Underlying Energy

The ecological to-do list for action is long at every level. It can seem overwhelming and hopeless. But as Pope John Paul II told us, love energizes—the love of the Creator and ours in return; our love of nature and the amazing animal and plant species so compelling in their beauty; our love for our own children, and all children; our love for those who are hungry and thirsty and don't deserve to live with such suffering; our love for people we haven't met in cultures across the world; our love of justice and kindness and heroism; our love for the vastness of the universe, and the vastness within us all; our love for the imperfections and messiness of life; our love of Christ and His for us. Pope John Paul II told young people:

> Another aspect of Christ's newness that you are to witness before the world: a new relationship with the environment surrounding you. Above all in this past century, people have made a use of the earthly realities which . . . has been irresponsible. *It is necessary to learn to look at nature with new eyes. . . .*
>
> Who can do so better than the Christian who is guided by faith to discover the work of the Creator in the realities of the world? Sun, stars, water, air, plants and animals are gifts with which God has made comfortable and beautiful the home prepared for people. *Whoever has understood this will treat them with responsible attention.*
>
> Having been redeemed by Christ, people can and must love the things that God has created and look at them as if they were coming from God's hands now. . . . The believer's attitude does not make a foolish use of resources or use arbitrary violence towards the animal kingdom to which all people belong in bodily dimension. . . .
>
> . . . be witnesses of the risen Christ in the newness of your personal life, in the newness of your relationship with others, in the newness of your relationship with the environment.
>
> ADDRESS TO YOUTH IN VITERBO, ITALY, MAY 27, 1984

To nurture (or discover) this new relationship with nature, we all have to get outdoors in it, without the distractions of screens and phones, so we can actually see it with new eyes, hear the birdsong and wind, the water running and frogs peeping. The Sabbath day, which the Bible states should be a full day of rest, is the perfect time to honor God with others in a house of worship, but also to spend time contemplating or just enjoying the multitude of divine expressions bursting forth in God's own natural cathedrals. As the pope told young people, get outdoors as often as you can to de-stress and to see God in nature:

It is good for people to read this wonderful book—the "book of
nature"—which lies open for each of us....

So my hope for you young people is that your "growth in statue
and in wisdom" will come about through contact with nature. Make
time for this! Do not miss it!

APOSTOLIC LETTER, *DILECTI AMICI* , MARCH 31, 1985

Aldo Leopold once said to his students:

If the individual has a warm personal understanding of the land, he
will perceive of his own accord that it is something more than a
breadbasket. He will see land as a community of which he is only a
member, albeit now the dominant one. He will see the beauty as well
as the utility of the whole, and know the two cannot be separated. We
love (and make intelligent use of) what we have come to understand.

"WHEREFORE WILDLIFE ECOLOGY?" *THE RIVER OF THE MOTHER OF GOD*

He also added "Once you learn how to read the land, I have no fear
what you will do to it, or with it. And I know many pleasant things it will
do to you."

Interfaith Ecological Efforts

These pleasant things that nature offers can become a centering point for
preservation efforts. So can a shared faith in a Creator. In 1990, Pope John
Paul II stated that one of the blessings and opportunities buried in the ecological crisis is that we will find and build solidarity beyond the borders of
faith, politics, and country:

Even men and women without any particular religious conviction,
but with an acute sense of their responsibilities for the common good,
recognize their obligation to contribute to the restoration of a healthy
environment. All the more should men and women who believe in
God the Creator, and who are thus convinced that there is a well-
defined unity and order in the world, feel called to address the problem.

Christians, in particular, realize that their responsibility within creation and their duty towards nature and the Creator are an essential part of their faith. As a result, they are conscious of a vast field of ecumenical and interreligious cooperation opening up before them.

Pope John Paul II was clearly not the lone religious leader calling for the care of creation and the ways of peace. Ecumenical Patriarch Bartholomew, with whom John Paul II wrote the joint declaration on the environment in 2002, and the 14th Dalai Lama, Tenzin Gyatso, immediately come to mind. The pope also worked with Archbishop Desmond Tutu, Lutheran Bishop James R. Crumley, Jr., the chief rabbis of Poland and Rome, among many others, and Muslim leaders, to stress the importance of a global common good.

Over the last twenty years, a heightened recognition of the need for working together for the care of God's creation has been sweeping through the world's religions, faiths, and spiritual groups.

In 2006 in the United States, nearly 90 evangelical leaders, rallied by Rich Cizik, then the Vice President for Governmental Affairs of the National Association of Evangelicals (NAE), drafted the Evangelical Climate Initiative. Signers included Rick Warren, pastor and author of *The Purpose-Driven Life*, and Leith Anderson, former NAE president. Cizik stated, "I don't think God is going to ask us how He created the Earth. But He will ask us what we did with what He created." The NAE would not officially sign it right away because it was considered controversial, but by the following year it signed, representing 45,000 churches and 30 million churchgoers. By 2011, more than 220 evangelical leaders had signed the call to action.

These are just the Evangelical Christians, a small drop in the new flow of denominations and faith communities across the spectrum who are finding their voice on these issues. Others are the Green Zionist Alliance, the Coalition on the Environment and Jewish Life, Jewcology, Hazon, Green Muslims, the Muslim Green Team, the Islamic Foundation for Ecology and Environmental Sciences, the Africa Muslim Environment Network (AMEN), Association of Buddhists for the Environment, the Mennonite

Creation Care Network, the Quaker Earthcare Witness, and so many more organizations in most every religion on earth.

The United Nations has an Interfaith Partnership for the Environment program in place since 1986. While world leaders met in Copenhagen in 2009 to talk about climate conventions and progress, over 6,000 members of different faiths met at the Parliament of the World Religions in Melbourne and sent statements encouraging the world leaders to commit to stronger standards to protect the planet.

Pope John Paul II stated how important interfaith efforts are for building human unity for good:

> ...the most pressing issues facing humanity—ecology, peace and the coexistence of different races and cultures, for instance—may possibly find a solution if there is a clear and honest collaboration between Christians and the followers of other religions and all those who, while not sharing a religious belief, have at heart the renewal of humanity.
>
> ENCYCLICAL, *FIDES ET RATIO*, SEPTEMBER 14, 1998

John Paul II's Lasting Interfaith Legacy

The pope's interfaith solidarity efforts brought real healing and paved the way for collaborations long after he died in 2005. Jewish and Muslim leaders in particular mourned his passing. After his death, a spring celebration was held at the Pope John Paul II Cultural Center (now the St. John Paul II Shrine) in Washington DC, in anticipation of what would have been John Paul the Great's 85th birthday. Imam Yahya Hendi, Muslim chaplain at Georgetown University, honored the late pope in a presentation stating what an extraordinary leader he was:

> He built bridges between religious communities and promoted religious freedom and tolerance. His passion for his ministry, compassion for the poor, and his respect for human life made him a beacon of light and a source of hope not only for Catholics but for all people, including Muslims....

...Politicians and leaders who seek temporary solutions to transient ideas should study this man's life, mission, and works. John Paul's message of world freedom and peace is his legacy to all of us...

PR NEWSWIRE, APRIL 7, 2005

In addition to apologizing for the Crusades and the Inquisition, Pope John Paul II recognized and highlighted the spirit of the Creator alive in the Muslim faith. "He emphasized the shared values of Catholics and Muslims, not their differences," said the imam.

At the same celebration, Rabbi Abie Ingber of the Hillel Jewish Student Center at the University of Cincinnati confirmed this and observed that:

> ...the Holy Father arguably has done more for interreligious relationships than any other person in history. By doing so, he has helped to set the world on a new path of mutual understanding...
>
> ...The Pontifex Maximus, the Master Bridge-Builder, has built a bridge of historic proportions. He was not afraid. We now must embrace his legacy and walk on the bridge together. Together we can find the spark of our common humanity...together we can create miracles; together on the narrow bridge that Karol Wojtyla, Pope John Paul II, built, we will not be afraid.

Pope Francis has followed this same course of bridge-building and peacemaking. As a cardinal, he wrote a book on economics with one of his rabbi friends; soon after becoming pope, he asked Church leaders to reach out to members of other faiths, particularly Islamic believers, and to countries that do not have official relations with the Holy See, such as China. He explained the word "pontiff" itself means builder of bridges:

> My wish is that the dialogue between us should help to build bridges connecting all people, in such a way that everyone can see in the other not an enemy, not a rival, but a brother or sister to be welcomed and embraced.

ADDRESS TO VATICAN DIPLOMATIC CORPS, MARCH 22, 2013

A Special Call to Roman Catholics
for Ecological Leadership and Action

For Pope John Paul II, the interfaith, intersecular movement for care of creation was a matter of life and death for humans as a species. As head of the Roman Catholic Church, though, *Il Papa* called specifically on the 1.6 billion believers to take urgent action. His spiritual command resounded:

> At the conclusion of this Message, I should like to address directly my brothers and sisters in the Catholic Church, in order to remind them of their serious obligation to care for all of creation. The commitment of believers to a healthy environment for everyone stems directly from their belief in God the Creator, from their recognition of the effects of original and personal sin, and from the certainty of having been redeemed by Christ. Respect for life and for the dignity of the human person extends also to the rest of creation, which is called to join man in praising God (cf. Ps. 148:96).
>
> WORLD DAY OF PEACE MESSAGE "PEACE WITH GOD THE
> CREATOR, PEACE WITH ALL CREATION," JANUARY 1, 1990

Coming at the end of a long speech about the ills needing to be addressed, the meaning was loud and clear. Catholics, GET TO WORK! Roman Catholics make up one sixth of the globe's population—and former Catholics a great portion more. Whenever and wherever the Catholic imagination of service becomes engaged in a good cause, an enormous amount of work gets done. Consider the numerous Catholic hospitals, hospices, schools for the poor, orphanages, and medical clinics that have been built in communities around the globe.

With such a cultural legacy of service, Catholics are *an untapped alternative energy.* They need to be called upon to apply their traditions of service and sacrifice to participate in ecological restoration, environmental activism, and sustainable development work—not just as individual humans and planetary citizens, but also specifically *as Catholics, working in faith.* Perhaps the Creation Care movement as a whole needs to put to use the old powerhouse of Catholic "guilt." Oft mocked, it's just the negative side of the positive call to responsibility and duty—abandoned concepts in our media-driven culture and even in many religious circles.

No "couch potato Christians" need apply. Pope Francis, echoing the voice of the great John Paul II, calls the Catholic Church to stop slumbering and get to work:

> We are in the time of action—the time in which we should bring God's gifts to fruition, not for ourselves but for him, for the Church, for others. The time to seek to increase goodness in the world always; and in particular, in this period of crisis, today, it is important not to turn in on ourselves, burying our own talent, our spiritual, intellectual, and material riches, everything that the Lord has given us, but, rather to open ourselves, to be supportive, to be attentive to others.
>
> In the square I have seen that there are many young people here …I ask you who are just setting out on your journey through life: have you thought about the talents that God has given you? Have you thought of how you can put them at the service of others? Do not bury your talents! Set your stakes on great ideals, the ideals that enlarge the heart, the ideals of service that make your talents fruitful. Life is not given to us to be jealously guarded for ourselves, but is given to us so that we may give it in turn.
>
> Dear young people, have a deep spirit! Do not be afraid to dream of great things!
>
> GENERAL AUDIENCE, APRIL 24, 2013

Both Francis and John Paul II have advised us to begin by taking the first steps, the ones nearest to us on the path, and never stop seeking the next steps, one task and one day at a time, without trying to shoulder the whole burden at once. Set ambitious goals—expect the long adventurous hike up the mountain—but do it together, and take regular rests to look around and enjoy the scenery. Bring as many young people as possible with you. John Paul II addressed those living in the mountains he so dearly loved, and talked of the powers and challenges of the earth's commanding physical presence:

So, dear friends...you are, as it were, molded by the mountain, by its beauty and its severity, by its mysteries and its attractions. The mountain opens its secrets only to those who have the courage to challenge it.

It demands sacrifice and training. It requires you to leave the security of the valleys but offers spectacular views from the summit to those who have the courage to climb it. Therefore it is a reality that strongly suggests the journey of the spirit, called to lift itself up from the earth to heaven, to meet God.

ANGELUS IN CAMPO IMPERATORE IN THE APENNINES, JUNE 20, 1993

(This message was delivered in a mountain pass area of Gran Sasso, in the Apennines. After the pope's death, on what would have been his 85th birthday (May 18, 2005), one of Gran Sasso's three peaks, the Gendarme, was renamed John Paul II Peak.)

Transforming Inflexible, Disrespectful, Angry Speech

To take these steps together toward restoring and preserving the planet's systems and species, we need a cultural change in everyday rhetoric. In his very first World Day of Peace address on January 1, 1979, Pope John Paul II reminded people of their duty to be civil:

Language is made for expressing the thoughts of the heart and for uniting. But when it is the prisoner of prefabricated formulas, in its turn it drags the heart along its own downward paths....

One must therefore act upon language in order to act upon the heart and avoid the pitfalls of language. It is easy to note to what an extent bitter irony and harsh judgments and criticism of others, especially "outsiders," argumentation, and an insistence on our own claims overrun our speech and strange social charity and justice itself.

By expressing everything in terms of relations of antagonistic force—of group and class struggles, of friends and enemies—an atmosphere is created for social barriers, contempt, even an underhanded or open support for hatred and terrorism....

> On the other hand, a heart devoted to the higher value of peace produces a desire to listen and understand, respect for others, gentleness which is real strength, and trust.
>
> Such a language puts one on the path of objectivity, truth and peace.
>
> WORLD DAY OF PEACE MESSAGE, "TO REACH PEACE, TEACH PEACE," JANUARY 1, 1979

The pope realized how daunting it is to look for solutions to ecological problems both locally and globally because there are so many different perspectives, approaches, needs and wants. And he followed his own counsel. With the Ecumenical Patriarch Bartholomew I he wrote:

> God has not abandoned the world. It is His will that His design and our hope for it will be realized through our cooperation in restoring its original harmony....
>
> Promote a peaceful approach to disagreement about how to live on this earth, about how to share it and use it, about what to change and what to leave unchanged.
>
> It is not our desire to evade controversy about the environment, for we trust in the capacity of human reason and the path of dialogue to reach agreement. We commit ourselves to respect the views of all who disagree with us, seeking solutions through open exchange, without resorting to oppression and domination.

John Paul II recognized that order to work on issues as big as climate change, ocean care, and reforestation, we will have to be able to talk to each other and listen, building relationships of trust and encouragement. Real scientific and social information will have to shared and agreed upon as truthful and "unbiased." From talk radio to TV shows, to music and videos, Facebook, tweets, and blogs, pulpits to podiums, the mockery, sarcasm, criticism, and hate speech will have to be replaced with dialogue, respect, and understanding. Presently things have gotten so caustic that the Y Generation has been dubbed by some bloggers "the Hater Generation." A change has to ripple throughout society, to transform global culture into one where no one gets voted off the island. The pope challenged media, sponsors, program hosts, policymakers, viewers, and writers to higher standards. Stories of courage, faith, heroism and solutions need to replace ones of cynicism, sex, death, and destruction:

> In this regard, the social communications on the media have a great educational task. The modes of expression in political debates, exchanges and confrontations, both national and international, are also influential. Leaders of the nations and of the international organizations must learn to find a new language, a language of peace: by its very self, it creates room for peace....
>
> Young people, be builders of peace. You are workers with a full share in producing this great common construction. Resist the easy ways out which lull you into sad mediocrity; resist the sterile violence....
>
> Take care that the legitimate desire to communicate ideas is exercised through persuasion and not through the pressure of threats and arms.
>
> By making resolute gestures of peace you will release the true aspirations of the peoples and will find in them powerful allies in working for the peaceful development of all.
>
> WORLD DAY OF PEACE MESSAGE, "TO REACH PEACE, TEACH PEACE," JANUARY 1, 1979

The Human Tipping Point?

After Pope John Paul II's death, Johan Vilhelm Eltvik, a Lutheran minister serving as the Secretary General of YMCA Europe, described the pope's part in the fall of communism in Europe:

> Sometimes the good Lord above blesses us with leadership bigger than we deserve... The spiritual authority of John Paul II was joined by people like Sakharov, Solzhenitsyn, Lech Walesa, and Vaclav Havel— just to mention a few—and they became stronger than any other authority, physical or ideological, and like the old walls in Jericho, the Berlin wall came tumbling down and our continent changed dramatically...
>
> SECRETARY GENERAL'S REPORT TO THE EUROPEAN ALLIANCE
> OF YMCAS, GENERAL ASSEMBLY LITOMYSL, CZECH REPUBLIC, MAY 5, 2005

This could happen again, only this time it could be the great wall of indifference and moral confusion toward nature will come tumbling down.

A spiritual and ecological consciousness is already rising from grassroots up, in every secular and faith group around the world. John Paul II's words might offer a different kind of centering point for regular Sunday preaching on themes of Creation Care in Scripture and everyday discussions on the relatedness of all life issues, for a new cross-cultural awakening and ecological solidarity. Consider the many dynamic and often heated and passionate debates that could go on around dinner tables and water coolers, at Scripture study groups and political forums, at book clubs and in university classrooms, driving people to deeper thinking and more responsible action.

What would happen if just 10% of Catholics and fallen-away Catholics could get vocally, physically, and spiritually engaged in conservation work and activism as individuals, and as parishes and dioceses, because of their reinvigorated faith? And what if they worked alongside other Christians, members of other faiths, secular activists, nonprofit staff, and everyday volunteers all in respectful solidarity for the same goals?

What if business and government leaders and media representatives around the world started feeling the pressure and the inspiration, and followed suit?

Imagine the momentum.

Think of how self-perpetuating, entrenched paradigms would bust apart and be tossed out. The culture of waste could be transformed to a culture of life, full-spectrum, all-inclusive, life.

New outlooks would be lively and surprising, revitalizing economies around the world with dynamic new technologies, products, approaches, goals, and sustainable patterns. That says nothing of what we could do to reduce terrorism, loss of diversity, and war, while improving our mental and physical health, prayer and spiritual life, relationships, and overall quality of life. Could John Paul II's call for urgent ecological action help catalyze the world, orienting cultures and nations toward care and action?

It could be the tipping point to converting our planet to a more spiritually and ecologically harmonious culture. And it could even be the gift that repairs and rebuilds the Catholic Church itself.

The Costly Price of Not Engaging with Youth

If people choose to turn away from their vocation and do not address the ecological problems as a matter of faith and morals, involving young people in the process, they will find themselves in spiritual and physical peril—not just because of an accounting with their Maker, not just because they will have neglected the needs of the poor, and not just because the economic systems will unravel as the natural resources do, but because the world's youth will turn away in cynical disgust. They will be disillusioned with the hypocrisy of faith if we pass on to them a world degraded and inhospitable to life because of our selfishness: our pollution, waste, exploitation, overharvesting, overdevelopment, and loss of species. They shall find religion irrelevant and impotent, and God's commands neglected.

Martin Luther King, Jr., wrote from jail about this same problem in the 1960s, and his words ring just as true today: "*Every day I meet young people whose disappointment with the church has turned into outright disgust.*"

He was addressing Christian churches about civil rights and justice, but his words apply to our present challenges:

> So often the contemporary church is a weak, ineffectual voice with an uncertain sound. So often it is an archdefender of the status quo. Far from being disturbed by the presence of the church, the power structure of the average community is consoled by the church's silent—and often even vocal—sanction of things as they are.
>
> But the judgment of God is upon the church as never before. If today's church does not recapture the sacrificial spirit of the early church, *it will lose its authenticity, forfeit the loyalty of millions, and be dismissed as an irrelevant social club with no meaning for the twentieth century....*

Because of the cases of child sexual abuse and cover-ups that followed, the Catholic Church has an even greater challenge restoring credibility and authenticity with youth than other religions. Pope Francis is reforming canonical law to make the Vatican and Church more transparent, responsible, and accountable. To rebuild trust with youth, however, something positive and life-giving, inspiring, and based in God must be put in the place of the betrayal. *Care for creation is this opportunity to switch things around and open the Church to the new movement of the Holy Spirit within it, to work with youth outdoors and among the less affluent on the missions they naturally care about.*

When talking to youth in Italy in 1985, John Paul II noted their disappointment in their elders and the world that they were passing on:

> In this situation, you young people can rightly ask the preceding generations: How have we come to this point? *Why have we reached such a degree of peril for humanity all over the world?* . . .
>
> *You young people can ask all these questions, indeed you must!* For this is the world you are living in today, and in which you will have to live tomorrow, when the older generation has passed on.

APOSTOLIC LETTER, DILECTI AMICI, MARCH 31, 1985

Artists and Musicians Needed to Unite the Culture and Pass on the Dream

Many youth feed on music and self-expression. We need to get them involved along with us in good, meaningful work to build a better ecological future. In considering how to bring about cultural solidarity for the planet and a springtime of the spirit and hope, the poet in Pope John Paul II understood the role of art and beauty to inspire. In his 1999 "Letter to Artists," the pope sent out a special plea to those skilled in the arts to utilize the Holy Spirit within themselves to bring to life the highest longings of humanity within the world and the Church: "Humanity in every age, and even today, looks to works of art to shed light upon its path and its destiny." He wrote:

> Those who perceive in themselves this kind of divine spark which is the artistic vocation—as poet, writer, sculptor, architect, musician, actor and so on—feel at the same time the obligation not to waste this talent but to develop it, in order to put it at the service of their neighbor and of humanity as a whole. . . .
>
> Society needs artists, just as it needs scientists, technicians, workers, professional people, witnesses of the faith, teachers, fathers and

mothers…Obedient to their inspiration in creating works both worthwhile and beautiful, they not only enrich the cultural heritage of each nation and of all humanity, but they also render an exceptional social service in favor of the common good.

Ecological flash mobs. Dancers weaving in and out of families in a park. Fountains for kids to cool themselves from the summer heat, with water spitting from whale and dolphin sculptures in a poor inner-city neighborhood. A youth orchestra playing Mozart with instruments made from materials picked from the trash, as shown in the movie *Landfill Harmonic*.

Art needs to burst out all over, to motivate all of us to go beyond exhaustion or despair or apathy to take care of this world, to slow down and savor life itself—each moment luring us out of our narrow notions and isolated worlds, enlarging our visions of the possible, engaging us all in the freeing, joyful life of creativity in our own spaces, and transforming lives. Art can embody the power of hope—fighting the toxic cynicism that is a contagious epidemic in the world today.

The Spiritual Benefits

As we change our orientation from despair to action and from "me" to "we," we begin to grow in compassion and living prayer. A priest at St. Patrick's Church in Chicago gave a very short Lenten sermon in 2013 in which he said:

> There are three words not used in the Lord's Prayer—"I," "me," and "mine." Jesus said "our," "we," and "us"—"*Our* Father…forgive *us our* trespasses as *we* forgive those…" We must remember how to pray and how to live. As Jesus did.

Pope John Paul II believed that we *can* become unified in solidarity toward the common good, that we can learn how to speak to each other about our deeper values and listen with our hearts, restructuring our protocols and policies, habits and products to fit God's designs in creation. That we can sacrifice excesses now to have plenty in the future. And if we do, we will be happier, healthier, and more economically stable, offering a better quality of life for all. We will reduce hostilities and conflicts, earning mutual respect as allies with shared values. We will reforest and regrass our globe, rebuilding its fertility and habitats for diverse species, including people.

As we integrate new, more harmonious forms of energies and technologies, we will spur new growth in our economies. Rich countries will find new markets by sharing technologies and resources with poorer ones, improving the health of the air, water, oceans, and land, and reducing the number of disaster refugees. Threatened species on the land and in the oceans will begin to multiply. As we each focus more on "being" than on "having," resting outdoors and with our families and friends, we'll enjoy life more while reducing our stress, materialism, waste, and pollution. As we heal nature, it will heal us.

Overall, John Paul II saw how all of us can more intimately become co-creators with God and live up to the responsibilities given to us in Genesis—to restore and maintain the Garden as a habitable planet for all, now and in the future. By working to restore and transform the planet, we can restore and transform ourselves—as individuals, as communities, as nations, and as a planet with new visions, new energies, new service, and new love, flowing from the source of all Love—the Creator, Son, and Holy Spirit. And the Catholic Church can also be transformed and renewed in the process.

But as the pope said, we can only accomplish this if we let love drive out fear in the face of all our many problems:

> In order to ensure that the new millennium now approaching will witness a new flourishing of the human spirit, . . . men and women must learn to conquer fear. *We must learn not to be afraid*, we must rediscover a spirit of hope and a spirit of trust.
>
> Hope is not empty optimism springing from a naive confidence that the future will necessarily be better than the past. Hope and trust are the premise of responsible activity and are nurtured in that inner sanctuary of conscience where "man is alone with God" and he thus perceives that he *is not alone* amid the enigmas of existence, for he is surrounded by the love of the Creator!

ADDRESS TO THE 50TH GENERAL ASSEMBLY OF THE
UNITED NATIONS, NEW YORK, OCTOBER 5, 1995

Life is a talent entrusted to us so
that we can transform it and increase it,
making it a gift to others.
No person is an iceberg
drifting on the ocean of history.
Each one of us belongs to a great family,
in which he has his own place and his own role to play.

Selfishness makes people deaf and dumb;
love opens eyes and hears, enabling people to make that original
and irreplaceable contribution which—together with the thousands
of deeds of so many brothers and sisters, often distant and unknown—
converges to form the mosaic of charity
which can change the tide of history.

11TH WORLD DAY OF YOUTH, NOVEMBER 25, 1995

Stories of Restoration and Hope from St. John Paul II

In the late 1980s, a young artist named Lily Yeh received a grant to build an art program in a poverty- and violence-stricken neighborhood of North Philadelphia. Thinking she would be starting her project in a small park, she arrived instead to an abandoned lot filled with broken glass and litter. Killing time while trying to figure out what to do, she started picking up little pieces of the trash that glittered. A few kids came around wondering what she was doing, and she got them involved. The trash they collected grew to become one sparkling mosaic sculpture, and then another, and then another.

Eventually, this vacant lot was transformed into a kind of art park, with mosaic cement trees and park benches. The mosaics and artworks expanded into other vacant lots nearby, but then the kids wanted *real* trees. Soon her program evolved the goal to re-green the area with gardens and an urban tree farm run by the neighborhood youth. They became the neighborhood green guardians who offered their harvests to the elderly poor.

Lily Yeh's Village of Arts and Humanities became a nonprofit organization in 1989. Since then it has blossomed into a community development program, sprouting dance and theater and music and visual arts projects for youth throughout the city. The kids involved use their artistic skills and joyful labor to combine art with nature, to bring back the beauty and health of life in the denuded Philadelphia neighborhoods.

The program has also been exported to sister neighborhoods in Kenya. Colorful murals of angels made by local kids out of broken bits of glass and plastic now overlook the garbage dumps where they pick up trash for a living to remind them of their heavenly worth. They are transformed from children who feel that they are much like the waste they collect, into artists and visionaries who see resources rather than trash.

This all began because one young woman started picking up a few pieces of shiny broken glass instead of walking away.

Pope John Paul II knew how easy is it to get discouraged and just walk away, but he told us to fight that despair with the knowledge of our own power to change the world:

> People sometimes have the impression that their individual decisions are without influence at the level of a country, the planet, or the cosmos. This gives rise to a certain indifference due to the irresponsible behavior of some individuals.
>
> However, we must remember that the Creator placed people in creation, commanding them to administer it for the good of all, making use of intelligence and reason. From this, we can be assured that *the slightest good act of a person has a mysterious impact on social transformation and shares in the growth of all. . . . everyone is invited to a profound personal conversion in their relationship with others and with nature.*
>
> This will enable a *collective conversion* to take place and lead to a life in harmony with creation. Prophetic actions, however slight, are an opportunity for a great number of people to ask themselves questions and to commit themselves to new paths.

ADDRESS TO PONTIFICAL ACADEMY SCIENCES "CONTRIBUTION
OF SCIENCE TO SOCIETY," MARCH 12, 1999

That is the faith that always guided St. John Paul II. To act each day as if your small choices will add up to big things. In the end, by joining with the choices of others, they do. Here are three stories of hope, responsible action, and restoration based on acts of John Paul II and others, that grew like the mustard seed into large trees bearing much fruit after he left.

Restoring the Human Family

It's October 1, 1979, and the brand new pope has landed at Logan International Airport in Boston, touching down on the soil of the United States for the first time as pontiff. Pope John Paul II is met by American Catholic cardinals and bishops, along with Senator Edward Kennedy and First Lady Rosalynn Carter. But more significantly, two hundred leaders from the Orthodox and Protestant religions meet him as well. He's known for his ecumenical and interfaith dialogue, and he's as popular as any rock star and more popular than any other world leader. Camera flashes sparkle in the air and throngs push forward to touch him. He offers a beaming wish for all Americans that "you fulfill completely your noble destiny in service to the world."

Forty thousand people have been waiting for hours to be present at his outdoor Mass on the Boston Common. On his way to Holy Cross Cathedral for a more intimate prayer service before going to the Common, the pope takes a fifty-five-minute scenic route through the many ethic neighborhoods of Boston—South Boston, Dorchester, North End, and South End—smiling and waving from an open-topped limousine. Tens of thousands line the boulevards. Confetti, streamers, and signs bid him welcome in Italian, Polish, Spanish, Greek, Arabic, Chinese, Vietnamese, English, and other languages as the mobs joyously wave back at him, chanting "Il Papa," or "Papa Juan Pablo," or "WE LOVE YOU!" in the misty drizzle. Citizens in Dorchester call out to him in Polish: *"Sto Lat Niech Zyje"—May you live a hundred years!*

Despite the United Nations of languages and faces that glow back at him, racial tensions are high. Three days earlier, an African American high school sophomore named Darryl Williams had been shot in the neck by three young white men at halftime as the star athlete played against a predominantly white school in Charlestown. It had only been five years since the city had implemented busing to desegregate schools, and the nearly all-black Jamaica Plains Knights had not been welcomed at the Charlestown football field. The players had been afraid to come. The mayor hadn't mince words. This was a "hate crime." Eighteen hundred of Williams's supporters marched to the cathedral, demanding justice.

In response, within the cathedral, the pope called the people of Boston to follow Christ, to act with their better selves and remember who they need to continue to be: "a community where people of all backgrounds, creeds, races, and convictions have provided workable solutions to problems and have created a home where all people can be respected in their human dignity." He stepped out to greet the protestors in compassion and empathy, and they met him with cheers rather than protests.

Later, at the Mass in the Commons, now in the pouring rain, the pope led a prayer petition that Darryl Williams live and have a complete recovery. At her own Congregational Church in Roxbury, his mother, Shirley Simmons, had earlier led others in similar prayers, begging those around her to fight the violence with prayers, not protests, riots, and revenge: "My nature is to ask for peace. . . . a lot of people were upset with me for asking for peace. But . . . there's a power greater than I am that will take care of this."

Violence and riots did not break out. Revenge was not the road. Shirley Simmons, Darryl Williams, and John Paul II, along with African American State Representative Mel King and other secular and spiritual leaders, helped move Boston forward, albeit slowly, toward more ethnic unity and justice, with solidarity against violence in the cause of human dignity.

Darryl Williams did not receive the full physical recovery that was prayed for, but he lived. And how he lived made all the difference. Though fighting for his life as a quadriplegic, he received racial death threats in his hospital room, where he was confined for months, unable to speak. But when two men from his neighborhood offered to take down some white people in revenge, Williams blinked "no." Lying in bed for days, months, and years, Williams prayed a lot.

He eventually chose to dedicate his life to promoting forgiveness, racial harmony, and equality, while also advocating for the handicapped. No one who met him was left untouched. After years of struggle, Williams gained fame as a motivational speaker and as an outreach sports specialist for Northeastern University's Center for the Study of Sport in Society, talking to young people and helping lead Project Teamwork, one of the nation's most successful violence-prevention programs. He became a spiritual athlete and hero, dubbed "America's Nelson Mandela."

In 2010, he died as a beloved son of Boston, acclaimed by everyone from President Bill Clinton and Senator Kennedy to Muhammad Ali. "Hate is a useless emotion that takes up too much energy," Williams said, and proved it (DAN SHAUGHNESSY, BOSTON GLOBE SUNDAY MAGAZINE, SEPTEMBER 20, 1990). In contrast, his three assailants moved in and out of the prison system because of this and other crimes. In time, the man who pulled the trigger on Darryl Williams died by gunshot himself, taken down by his cousin.

The Recycled Park

On the other side of Boston, in Dorchester, where the pope was saluted in Polish, another story of restoration and solidarity was developing. Dorchester is bordered on the east by Dorchester Bay and to the south by the estuary and salt marshes of the Neponset (or Harvest River, in the Massachusetts Indian language). Though much of the estuary lands and salt marshes had been set aside in the 1890s for preservation, during the 1940s they were drained, filled in, and turned into a municipal landfill and trash incinerator site, plus a drive-in and lumberyard. The river became so heavily polluted with industrial chemicals and sewage that swimming in or drinking the water were unthinkable. Some people tried to get things cleaned up, but it was an uphill battle.

In 1972, though, the Clean Water Act finally passed, and change suddenly seemed possible. Industries and cities had to stop polluting, or get sued. In 1977, a group was organized that morphed into the Boston Natural Areas Network (BNAN), "dedicated to bringing together local residents, partner organizations, public officials, and foundations to preserve, expand, and enhance urban open space, including community gardens, greenways, and urban wilds."

By the time the pope arrived in 1979, the drive-in had closed and been purchased by the Massachusetts Department of Conservation and Recreation, and the incinerator and landfill had shut down. Land from the closed lumberyard had also been acquired. Through work from the federal government, state, city, local activists, and businesses, the river gradually grew cleaner.

The city of Dorchester was a mixed culture and recycled place itself, having gone through many transformations over the centuries. Besides the native, colonial and African American communities, immigrants of all kinds came to Dorchester's shores and built up American families—the Irish and Polish, of course, but also other Eastern and Western Europeans, including Jews, and more recent immigrants from Puerto Rico, the Dominican Republic, Haiti, Jamaica, Trinidad and Tobago, Vietnam, Cape Verde, and other Latin American, Asian, and African nations. Many academics had also settled there. In addition, active professional, artistic, and gay neighborhood groups had coalesced. Overall, Dorchester was a hearty stew of God's people from diverse cultures and religions, heavily spiced with Catholics and home to four campuses of the archdiocese's Pope John Paul II Catholic Academy.

Around the time that the pope gave his "Peace with God the Creator, Peace with All Creation" speech in 1990, local Dorchester residents started working with the BNAN, state, and other partners on the idea of restoring the area for a mixed-use, "recycled" park. The diverse community came together to raise millions of dollars and cadres of volunteers and partners for this ecological restoration project.

Over the decades, four feet of clay were brought in to cover the landfill, and native grasses, reeds, trees, and shrubs were planted to create an open meadow and wetland savannah for birds and butterflies, hikes, and bike rides. Snowy egrets and blue herons returned along with teal and mergansers. Over the old lumberyards, plans for community gardens began sprouting. The former drive-in land was spread out smoothly for numerous soccer fields, so Dorchester could pursue with fervor the pope's favorite team sport (one at which he excelled in his youth, playing with his Jewish teammates). There would be a playground for families, picnic spots, and a boat dock.

In 2001, in honor of the pope's visit and what he had done for the world, they named the park after him, opening the first section of the sixty-five-acre park, with the rest following the next year. A brochure with a small wildlife-spotting game and guide for kids, "The Recycled Park: The Environmental Evolution of Pope John Paul II Park," tells what is there:

If you take a family bike ride through Pope John Paul II Park, you'll see much of the same plants and wildlife enjoyed hundreds of years ago along the "Neunsep." The restored salt marsh with its reed and cord grasses and salt marsh hay once again attracts egrets, kingfishers, herons and other shorebirds. If you look carefully, you'll catch a glimpse of the blue fish and striped bass that run the river again, and the many small creatures, such as toads, salamanders, the white-footed mouse and mayflies that make the wetland their home.

Pope John Paul II Park would, of course, be the perfect site for an annual Creation Sunday Mass in the spring, in addition to a fall interfaith St. Francis Feast Day service with blessings of the animals. Or simply the perfect site for Sabbath day activities. No culture of waste here. It has been transformed into one of life, in the spirit of Saints Francis and John Paul II. He once said: "Every kind of life should be respected, fostered, and indeed loved, as a creation of the Lord God, who created everything 'good.'" And here, in this park, it is (ADDRESS TO PONTIFICAL ACADEMY OF SCIENCES, STUDY WEEK ON MAN AND HIS ENVIRONMENT, MAY 18, 1990).

Hope at the Edge of the World

In Antarctica, on the northern edge of Hero Bay, a peninsula juts out from Livingston Island (one of the South Shetland Islands). It tells another story of ecological solidarity. Hero Bay is one of the sites where scientists from varied nations monitor the planet's climactic and ocean health. The stunning ridged, glacier-covered landmass that rises up alongside its waters has been named the Ioannes Paulus II Peninsula to honor the pope's passionate global leadership to build peace, foster understanding between nations and religions, and protect the planet.

At the end of the peninsula is a marine sanctuary for monitoring the living heritage of the sea and an Important Bird Area (IBA). These reserves attract the largest population of Antarctic fur seals in the region and a rookery of ten thousand pairs of Chinstrap penguins, as well as Gentoo Penguins, Antarctic terns, brown skuas, black-bellied storm petrels and cape petrels, kelp gulls, snowy sheathbills and imperial shags. How fitting that such an abundance of life hangs out on the peninsula named after this late, great holy man.

This peninsula park stands as a global measuring stick of how well we are caring for the planet. As St. John Paul dearly loved snow and skiing, it is our job to ensure that there are glaciers and snow remaining on those heights for generations to come.

This monumental job will take everyone's participation. It may be our snow-boarding youth who will display the energy, idealism, service, and new ideas needed to save that snow and ice. In the spring of 2009, Paul Hawken gave a commencement speech at the University of Portland that rings with truth for all. He entitled it, "You Are Brilliant, and the Earth is Hiring." Here are snippets of what he said:

> Basically, civilization needs a new operating system, you are the programmers, and we need it within a few decades.
>
> This planet came with a set of instructions, but we seem to have misplaced them. Important rules like don't poison the water, soil, or air, don't let the earth get overcrowded, and don't touch the thermostat have been broken....
>
> If you look at the science about what is happening on earth and aren't pessimistic, you don't understand the data. But if you meet the people who are working to restore this earth and the lives of the poor, and you aren't optimistic, you haven't got a pulse.
>
> What I see everywhere in the world are ordinary people willing to confront despair, power, and incalculable odds in order to restore some semblance of grace, justice, and beauty to this world....
>
> Inspiration is not garnered from the litanies of what may befall us; it resides in humanity's willingness to restore, redress, reform, rebuild, recover, reimagine, and reconsider...
>
> Nature beckons you to be on her side. You couldn't ask for a better boss... This is your century. Take it and run as if your life depends on it.

There is enough work for us all, and we need to encourage each other, young and old, to answer the call. St. John Paul II told the youth, and all of us with them, to be the world's missionaries of hope:

> Dear young people, love life! Love it with the depth and passion of Francis of Assisi. *Love it in the beauty of nature, in the joy of friendships, in the advances of science, in the generous struggle to build a better world.*
>
> ADDRESS TO YOUTH IN LA VERNA, ITALY, SEPTEMBER 17, 1993

Epilogue

St. Peter's Square
Tuesday, March 19, 2013
Solemnity of St. Joseph

Dear Brothers and Sisters,

I thank the Lord that I can celebrate this Holy Mass for the inauguration of my Petrine ministry on the solemnity of St. Joseph, the spouse of the Virgin Mary and the patron of the universal Church. It is a significant coincidence, and it is also the name-day of my venerable predecessor: we are close to him with our prayers, full of affection and gratitude...

In the Gospel we heard that "Joseph did as the angel of the Lord commanded him and took Mary as his wife" (Mt. 1:24). These words already point to the mission which God entrusts to Joseph: he is to be the *custos*, the protector. The protector of whom? Of Mary and Jesus; but this protection is then extended to the Church, as Blessed John Paul II pointed out: "Just as St. Joseph took loving care of Mary and gladly dedicated himself to Jesus Christ's upbringing, he likewise watches over and protects Christ's Mystical Body, the Church, of which the Virgin Mary is the exemplar and model."

How does Joseph exercise his role as protector? Discreetly, humbly and silently, but with an unfailing presence and utter fidelity, even when he finds it hard to understand. From the time of his betrothal to Mary until the finding of the twelve-year-old Jesus in the Temple of Jerusalem, he is there at every moment with loving care. As the spouse of Mary, he is at her side in good times and bad, on the journey to Bethlehem for the census and in the anxious and joyful hours when she gave birth; amid the drama of the flight into Egypt and during the frantic search for their child in the Temple; and later in the day-to-day life of the home of Nazareth, in the workshop where he taught his trade to Jesus.

How does Joseph respond to his calling to be the protector of Mary, Jesus and the Church? By being constantly attentive to God, open to the signs of God's presence and receptive to God's plans, and not simply to his own. This is what God asked of David, as we heard in the first reading. God does not want a house built by men, but faithfulness to his word, to his plan. It is God himself who builds the house, but from living stones sealed by his Spirit. Joseph is a "protector" because he is able to hear God's voice and be guided by his will; and for this reason he is all the more sensitive to the persons entrusted to his safekeeping. He can look at things realistically, he is in touch with his surroundings, he can make truly wise decisions. In him, dear friends, we learn how to respond to God's call, readily and willingly, but we also see the core of the Christian vocation, which is Christ! Let us protect Christ in our lives, so that we can protect others, so that we can protect creation!

The vocation of being a "protector," however, is not just something involving us Christians alone; it also has a prior dimension which is simply human, involving everyone. It means protecting all creation, the beauty of the created world, as the Book of Genesis tells us and as St. Francis of Assisi showed us. It means respecting each of God's creatures and respecting the environment in which we live. It means protecting people, showing loving concern for each and every person, especially children, the elderly, those in need, who are often the last we think about. It means caring for one another in our families: husbands and wives first protect one another, and then, as parents, they care for their children, and children themselves, in time, protect their parents. It means building sincere friendships in which we protect one another in trust, respect, and goodness. In the end, everything has been entrusted to our protection, and all of us are responsible for it. Be protectors of God's gifts!

Whenever human beings fail to live up to this responsibility, whenever we fail to care for creation and for our brothers and sisters, the way is opened to destruction and hearts are hardened. Tragically, in every period of history there are "Herods" who plot death, wreak havoc, and mar the countenance of men and women.

Please, I would like to ask all those who have positions of responsibility in economic, political and social life, and all men and women of goodwill: let us be "protectors" of creation, protectors of God's plan inscribed in nature, protectors of one another and of the environment. Let us not allow omens of destruction and death to accompany the advance of this world! But to be "protectors", we also have to keep watch over ourselves! Let us not forget that hatred, envy and pride defile our lives! Being protectors, then, also means keeping watch over our emotions, over our hearts, because they are the seat of good and evil intentions: intentions that build up and tear down! We must not be afraid of goodness or even tenderness!

Here I would add one more thing: caring, protecting, demands goodness, it calls for a certain tenderness. In the Gospels, St. Joseph appears as a strong and courageous man, a working man, yet in his heart we see great tenderness, which is not the virtue of the weak but rather a sign of strength of spirit and a capacity for concern, for compassion, for genuine openness to others, for love. We must not be afraid of goodness, of tenderness!

Today, together with the feast of St. Joseph, we are celebrating the beginning of the ministry of the new Bishop of Rome, the Successor of Peter, which also involves a certain power. Certainly, Jesus Christ conferred power upon Peter, but what sort of power was it? Jesus' three questions to Peter about love are followed by three commands: feed my lambs, feed my sheep. Let us never forget that authentic power is service, and that the Pope too, when exercising power, must enter ever more fully into that service which has its radiant culmination on the Cross.

He must be inspired by the lowly, concrete and faithful service which marked St. Joseph and, like him, he must open his arms to protect all of God's people and embrace with tender affection the whole of humanity, especially the poorest, the weakest, the least important, those whom Matthew lists in the final judgment on love: the hungry, the thirsty, the stranger, the naked, the sick and those in prison (cf. Mt. 25:31–46). Only those who serve with love are able to protect!

In the second reading, St. Paul speaks of Abraham, who, "hoping against hope, believed" (Rom. 4:18). Hoping against hope! Today too, amid so much darkness, we need to see the light of hope and to be men and women who bring hope to others. To protect creation, to protect every man and every woman, to look upon them with tenderness and love, is to open up a horizon of hope; it is to let a shaft of light break through the heavy clouds; it is to bring the warmth of hope! For believers, for us Christians, like Abraham, like St. Joseph, the hope that we bring is set against the horizon of God, which has opened up before us in Christ. It is a hope built on the rock which is God.

To protect Jesus with Mary, to protect the whole of creation, to protect each person, especially the poorest, to protect ourselves: this is a service that the Bishop of Rome is called to carry out, yet one to which all of us are called, so that the star of hope will shine brightly. Let us protect with love all that God has given us!

I implore the intercession of the Virgin Mary, St. Joseph, Sts. Peter and Paul, and St. Francis, that the Holy Spirit may accompany my ministry, and I ask all of you to pray for me! Amen.

PART V

RESOURCES AND SPEECHES

This very goal [of rediscovering our fraternity with the earth]
was foreshadowed by the Old Testament in the Hebrew Jubilee,
when the earth rested and man gathered
what the land spontaneously offered (Lev 25:11-12).
If nature is not violated and humiliated
it returns to being the sister of humanity.

GENERAL AUDIENCE, ZENIT TRANSLATION, JANUARY 26, 2000

Action Resources

It can be overwhelming and confusing to know where to start in more fully living up to our ecological vocations and duties, and to grow spiritually in the process. The following organizations have an incredible amount of guidance, actions, study materials, and events to offer, giving opportunities for personal reflection, prayer, and transformation, and group awareness and actions.

Materials and links are also posted on the *Following St. Francis* website: www.followingstfrancis.com

Roman Catholic
St. Francis Pledge: www.catholicclimatecovenant.org/the-st-francis-pledge
Franciscan Action Network (FAN): www.franciscanaction.org
Franciscan Earth Corps: www.franciscanaction.org/earthcorps
Catholic Climate Covenant: www.catholicclimatecovenant.org
U.S. Conference of Catholic Bishops Prayers to Care for Creation:
    www.usccb.org/prayer-and-worship/prayers/prayers-to-care-for-creation.cfm
St. Katerie Tekakwitha Conservation Center: www.conservation.catholic.org
National Catholic Rural Life Conference: www.ncrlc.com
Eagle Eye Ministries, from the Community of St. John: eagleeyeministries.org/
    about/about-eem

Christian
Blessed Earth: Serving God, Saving the Planet: www.blessedearth.org
Evangelical Environmental Network: www.creationcare.org
Creation Care Study Program: Christian Environmental Study Abroad: www.creationcsp.org
Restoring Eden: Christians for Environmental Stewardship:
    www.restoringeden.org
Renewal: Students Caring for Creation: www.renewingcreation.org
Center for Environmental Leadership: www.center4eleadership.org
Christians for the Mountains: www.christiansforthemountains.org
Earth Ministry: Caring for All Creation: www.earthministry.org
Eco-Justice Ministries: www.eco-justice.org

Interfaith Creation Care Organizations
National Religious Coalition on Creation Care: www.nrccc.org
Interfaith Power & Light: A Religious Response to Global Warming:
    www.interfaithpowerandlight.org

Cool Congregations: www.coolcongregations.org
Interfaith Ocean Ethic Campaign: www.oceanethicscampaign.org
Carbon Confession: Faith in Action: www.carbonconfession.com
ARC: Alliance of Religions and Conservation: www.arcworld.org
GreenFaith: Interfaith Partners for the Environment: www.greenfaith.org
The Regeneration Project: Deepening the Connection Between Ecology and
    Faith: www.theregenerationproject.org

Specific Non-Christian Faiths and Creation Care
Green Zionist Alliance: www.greenzionism.org
Coalition on the Environment and Jewish Life (COEJL): Protecting Creation
    Generation to Generation: www.coejl.org
Green Muslims: www.greenmuslims.org
Quaker Earthcare Witness: www.quakerearthcare.org
Buddhist Peace Fellowship:
    www.buddhistpeacefellowship.org/category/eco-justice
Indigenous Women's Network: www.indigenouswomen.org
Honor the Earth: www.honorearth.org
Indigenous Environmental Network: www.ienearth.org

Restoration Organizations
Green Belt Movement: www.greenbeltmovement.org
Savory Institute: www.savoryinstitute.com
Restoration Agriculture Institute: www.restorationag.org
Ocean Conservancy: www.millenniumvillages.org
Marine Stewardship Council: www.msc.org
Ciudad Saludable: www.ciudadsaludable.org
Sustainable South Bronx: www.ssbx.org
Village of Arts and Humanities: www.villagearts.org
Habitat for Humanity: www.habitat.org
Heifer International: www.heifer.org
Rocky Mountain Institute: www.rmi.org
Youth Conservation Corps: www.youthconservationcorps.org
Partners in Health: www.pih.org
Millennium Villages Project: www.millenniumvillages.org
A Rocha: Restoring People and Places: www.arocha-usa.org
Transition Town Network: www.transitionnetwork.org
Transition US: www.transitionus.org
Zero Waste Alliance: www.zerowaste.org

Major Papal Works on Creation Care, Ecology, and the Environment

Assembled below are some key works in English (with a few in Italian) that address or weave in ecological teachings. There are others in English and many more in other languages, especially Italian. To see most works in their entirety, visit www. vatican.va ("Addresses" are listed under Speeches).

For up-to-date additions, see the Catholic Climate Covenant and the Franciscan Action Network (as well as the Following St. Francis book website). For a summation of early papal statements, read the pioneering book by Sister Marjorie Keenan, RSHM, *From Stockholm to Johannesburg: A Historical Overview of the Concern of the Holy See for the Environment*, published by the Vatican Press in 2002.

Another great source of information and of popes' and saints' quotations on ecology is www.conservation.catholic.org, run by the St. Kateri Tekawitha Center.

Pope Paul VI
(Cardinal Karol Wojtyła worked with him on some of these documents)
Dogmatic Constitution: Lumen Gentium, 1964
Pastoral Constitution: Gaudium Et Spes, 1965
Encyclical: Populorum Progressio, 1967
"Justice in the World," Second Synod of Bishops, 1971
Apostolic Letter: Octogesima Adveniens, 1971
Message to the Stockholm Conference, 1972

Pope John Paul II
Message: "To Reach Peace, Teach Peace," World Day of Peace, January 1, 1979
Homily: Living History Farms, Des Moines, Iowa, October 4, 1979
Apostolic Letter: Inter Sanctos, November 29, 1979
Encyclical: Redemptor Hominis, 1979
Address: Appeal for the Drought-Stricken Sahel, Airport, Ouagadougou, Burkina Faso, May 1980
Address: Pontifical Academy of Sciences, Study Week on Mankind and Energy, November 14, 1980
Encyclical: Laborem Exercens, September 14, 1981
Address: Pontifical Academy of Sciences, Study Week on Cosmology and Fundamental Physics, October 3, 1981
Reconciliation Address: Interfaith Gathering at Assisi, March 3, 1982

Angelus: National Day of Ecology and Zoology, March 28, 1982 (in Italian)

Greeting: Participants in Terra Mater Seminar, Gubbio, October 3, 1982 (In Italian)

Address: Pontifical Academy of Sciences, Study Week on Biological Experimentation, October 23, 1982

Address: XII International Symposium on Nuclear Physics, December 1982

Address: Representatives of the World of Science and Art, Vienna, September 1983

Address: Pontifical Academy of Sciences, Study Week on Chemical Events in the Atmosphere and Their Impact on the Environment, and Science at the Service of Peace, November 12, 1983

Message: Radio and TV to Native Peoples of Canada, Yellow Knife, September 18, 1984

Address: Pontifical Academy of Sciences, Study Week on the Impact of Space Exploration on Mankind, October 21, 1984

Address: Vatican Observatory Conference on Cosmology, July 1985

Apostolic Letter: Dilecti Amici, (on the inauguration of the International Youth Year), March 30, 1985

Address: Members of the Agency of the United Nations Program on the Environment, Nairobi, Kenya, August 18, 1985

Address: FOA, XXXIII World Food Conference, November 10, 1985

Post-General Audience Greetings: World Environment Day, June 5, 1986

Address: to the Aborigines and Torres Strait Islanders, Alice Springs, Australia, November 29, 1986

Address: National Tekawitha Conference, Phoenix, September 14, 1987

Homily: Mass for Rural Workers in Laguna Seca, Monterey Peninsula, California, September 17, 1987

Homily: Mass for Native Peoples of Canada, Fort Simpson, September 20, 1987

Address: Pontifical Academy of Sciences, Study Week on Remote Sensing and its Impact on Developing Countries, June 20, 1986

Address: Pontifical Academy of Sciences, Study Week on Persistent Meteo-Oceanographic Anomalies and Teleconnections, September 26, 1986

Address: Plenary Assembly of Pontifical Academy of Sciences on the 50th Anniversary of its Foundation, October 28, 1986

Address: Pontifical Academy of Sciences, Study Group on A Modern Approach to the Protection of the Environment, November 6, 1987

Encyclical: Sollicitudo Rei Socialis, December 30, 1987

Address: XXIV World Food Conference, November 13, 1987

Address: Youth Movement of Italian National Farmers Conference, January 9, 1988

Address: Representatives of Science, Art, and Journalism, Festzpielhaus, Salzburg, Austria, June 26, 1988

Address: European Council, Strasbourg, France, October 8, 1988

Apostolic Exhortation: Christifideles Laici, December 30, 1988

Address: Pontifical Academy of Sciences, Study Week on Society for Development in a Solidarity Framework, October 27, 1989

Address: Participants in the Symposium Sponsored by Nova Spes International Foundation., December 1989

Address: XXV Session of the Conference of the United Nations Food and Agricultural Organization (FAO), November 16,1989

Message: "Peace with God the Creator, Peace with All Creation," World Day of Peace, January 1, 1990

General Audience: "The Creative Actions of the Divine Spirit," January 7, 1990

Address: Regional Council of Lazio, Rome, February 5, 1990

Address: Business Community, Durango, Mexico, May 10, 1990

Address: Pontifical Academy of Sciences, Study Week on Man and His Environment, Tropical Forests, and Conservation of Species, May 18,1990

Address: Pontifical Academy of Sciences, Study Week on Science in the Context of Human Culture I, October 29, 1990

Encyclical: Centesimus Annus (On the Hundredth Anniversary of Rerum Novarum), May 1,1991

Address: Young People in the Dolomites, June 28, 1991

Address: Pontifical Academy of Sciences, Study Week on in the Context of Human Culture II, October 4, 1991

Address: Pontifical Academy of Sciences, Study Week on Resources and Population, November 22, 1991

Message to Indigenous Peoples, Santo Domingo, Dominican Republic, October 12, 1992

Angelus: Campo Imperatore, June 20, 1993

Angelus: Santo Stefano di Cadore, July 11, 1993

Address: Pontifical Academy of Sciences, Study Week on Chemical Hazards in Developing Countries, October 22, 1993

Homily: World Youth Day, Cherry Creek Campground, Denver, August 14, 1993

Address: Pontifical Academy of Sciences, Study Week on Human Genome Project, November 20, 1993

Apostolic Letter: Tertio Millenio Adveniente, 1994

Address: Plenary Session, Pontifical Academy of Sciences, Human Genome; Alternative Energies for Developing Countries, et al, October 28, 1994

Address: Pontifical Academy of Sciences, Study Week on Scientific Bases of the Natural Regulation of Fertility and Associated Problems, November 18, 1994

Encyclical: Evangelium Vitae, March 25,1995

Letter to Women, June 19, 1995

Address: XXVIII Conference of the UN Food and Agricultural Organization (FAO), October 23, 1995

Address: 11th World Day of Youth, November 25, 1995

Address: Plenary Session, Pontifical Academy of Sciences. The Origins and Early Evolution of Life, October 22, 1996

Lenten Meditation: Season of Lent Offers Profound Lesson to Respect Nature, March 25, 1996

Address: European Bureau for the Environment, June 1996

Address: International Conference on Space Research, January 1997

Address: Rotary Club, Security and the Environment, Vatican City, March 24, 1997

Address: Conference on Environment and Health, Catholic University of Sacred Heart, March 24, 1997

Apostolic Letter: *Dies Domini*, July 5, 1998

General Audience: "The Holy Spirit Acts in All Creation and History," August 12, 1998

General Audience: "Creation Must Be Dwelling Place of Peace," August 19, 1998

Encyclical: *Fides et Ratio*, September 14, 1998

Address: Pontifical Academy of Sciences on Changing Concepts of Nature, October 27, 1998

Bull Indiction for Jubilee Year of 2000: *Incarnationis Mysterium*, November 20, 1998

Address: German Delegation with Christmas Tree, December 19, 1990

Message: "Respect for Human Rights: The Secret of True Peace," World Day of Peace, January 1, 1999

Homily: "Seek a Culture of Life, Not Death," St. Louis, Missouri, January 27, 1999

Address: Pontifical Academy for Sciences, Study Week, Science for Survival and Sustainable Development, March 12, 1999

Address: The World Federation of Scientists, March 1999

Letter to Artists, April 4, 1999

Address: Liturgy of the Word: "Creation and Humanity Threatened by Lack of Respect for Life," Zamość, Poland, June 1999

Post-Synodal Apostolic Exhortations: *Ecclesia in America, Ecclesia in Asia, Ecclesia in Oceania*, 1999

Message: "Peace on Earth to Those Whom God Loves," World Day of Peace, January 1, 2000

Message: To the UN Food and Agricultural Organization, World Food Day, 2000

General Audience: "Listening to the Word and the Spirit in Cosmic Revelation," August 2, 2000

Message: XXI World Day of Tourism, September 2000

Addresses: Jubilee of the Agricultural World, November 11–12, 2000

Address: Plenary Session, Pontifical Academy of Sciences, Science and Future of Mankind, November 13, 2000

Apostolic Letter: *Novo Millennio Ineunte,* January 6, 2001

Address: To the Diplomatic Corps, January 13, 2001

General Audience: "Call to Ecological Conversion," January 17, 2001

Address: On the Fifteenth Anniversary of the Chernobyl Nuclear Accident, Vatican City, April 27, 2001

Address: 16th World Youth Day, Assisi, Italy, August 26, 2001

General Audience: "Creation Is a Sacred Book, " January 30, 2002

Common Declaration on Environmental Ethics, Pope John Paul II and the Ecumenical Patriarch Bartholomew I, Rome-Venice, June 10, 2002

Address: To the Diplomatic Corps, January 10, 2002

Message: "Savage Tourism," for XXXIII World Day of Tourism, June 24, 2002

Address: XX World Food Day, FOA Food Summit, June 10, 2002

Angelus: Castel Gandolfo, August 25, 2002

Address: Plenary Session, Pontifical Academy of Sciences, The Cultural Values of Science, November 11, 2002

Message: Ecumenical Patriarch Bartholomew I, Fifth Symposium of the Religion, Science and the Environment Project, May 27, 2003

Address: Plenary Session, Pontifical Academy of Sciences, Paths of Discovery, November 2004

Message: "Do Not Be Overcome by Evil, But Overcome Evil with Good," World Day of Peace, January 1, 2005

Pope Benedict XVI

Homily: Inaugural Mass, April 24, 2005

Message: "The Human Person, the Heart of Peace," World Day of Peace, January 1, 2007

Homily: "Sabbath Rest and Creation," September 9, 2007

Question and Answer Session: With Clergy of Dioceses of Belluno-Feltre and Treviso, "Be Obedient to the Voice of the Earth," Church of St. Justin Martyr, Auronzo di Cadore, July 24, 2007

Address: To His Holiness Bartholomew I and Environmental Conference, "Highly Industrialized Nations Must Share Clean Technologies," September 1, 2007

Address: to Young People of Loreto, Italy, September 2, 2007

Angelus: On the Twentieth Anniversary of Montreal Protocol, September 16, 2007

Message: "The Human Family: A Community of Peace," World Day of Peace January 2, 2008

Address: People of Australia and Youth, World Youth Day, July 4, 2008

Discussion: Journalists on Plane, "The Gift of the Earth," July 14, 2008

Address: XXXIII World Youth Day, Sydney Australia, July 17, 2008

Address: Clergy of the Diocese of Bolzano-Bressanone, "Inseparability of Redemption and Creation," August 6, 2008

Annual Address to Curia: December 22, 2008

Encyclical Letter: *Caritas in Veritate*, June 29, 2009

General Audience: August 26, 2009

Message: Ecumenical Patriarch Bartholomeos I, Eighth Symposium of the
    Religion, Science and the Environment Project, October 12, 2009

Message: "If You Want to Cultivate Peace, Protect Creation," World Day of Peace
    January 2, 2010

Address: To the Diplomatic Corps, January 11, 2010

Message: 30th World Food Day, October 15, 2010

Message: "Educating Young People in Justice and Peace," World Day of Peace,
    January 1, 2012

Address: To the Diplomatic Corps, January 9, 2012

Pope Francis (through November 2013)

Comments to members of communication media on his choice of name, March
    16, 2013

Homily: Inaugural Mass, March 19, 2013

Address: New Pontifical Ambassadors, May 16, 2013

Message: World Environment Day, June 5, 2013

Message: To Bishops of Latin American and the Caribbean, "Protect Amazonia and
    Environment," July 27, 2013

Message: World Youth Day, July 28, 2013

Address: Meeting with Workers, Largo Carlo Felice, Cagliari, Sardinia, September
    22, 2013

Interview: Antonio Spadero, S.J., (held in 3 meetings in August 2012, Rome) trans-
    lated into English, published in America, September 30, 2003

Interview: Eugenio Scalfari, La Repubblica, October 1, 2013

Apostolic Exhortation: Evangelii Gaudium, November 24, 2013

Peace with God the Creator, Peace with All of Creation

January 1, 1990
POPE JOHN PAUL II'S WORLD DAY OF PEACE MESSAGE

This message, the most famous of Pope John Paul II's ecological addresses, sets forth most of the significant themes found in his many other works. It conveys the real urgency and significance of his call to faith, morality, and action, and the responsibilities of all to get to work and build solidarity for the common good.

In our day, there is a growing awareness that world peace is threatened not only by the arms race, regional conflicts and continued injustices among peoples and nations, but also by a lack of *due respect for nature*, by the plundering of natural resources and by a progressive decline in the quality of life. The sense of precariousness and insecurity that such a situation engenders is a seedbed for collective selfishness, disregard for others and dishonesty.

Faced with the widespread destruction of the environment, people everywhere are coming to understand that we cannot continue to use the goods of the earth as we have in the past. The public in general as well as political leaders are concerned about this problem, and experts from a wide range of disciplines are studying its causes. Moreover, a new *ecological awareness* is beginning to emerge which, rather than being downplayed, ought to be encouraged to develop into concrete programmes and initiatives.

Many ethical values, fundamental to the development of a *peaceful society*, are particularly relevant to the ecological question. The fact that many challenges facing the world today are interdependent confirms the need for carefully coordinated solutions based on a morally coherent world view.

For Christians, such a world view is grounded in religious convictions drawn from Revelation. That is why I should like to begin this Message with a reflection on the biblical account of creation. I would hope that even those who do not share these same beliefs will find in these pages a common ground for reflection and action.

I. "And God Saw That It Was Good"

In the Book of Genesis, where we find God's first self-revelation to humanity (Gen. 1–3), there is a recurring refrain: *"And God saw that it was good."* After creating the heavens, the sea, the earth and all it contains, God created man and woman. At this point the refrain changes markedly: "And God saw everything that he had made, and behold, *it was very good* (Gen. 1:31). God entrusted the whole of creation to the man and woman, and only then—as we read—could he rest "from all his work" (Gen. 2:3).

Adam and Eve's call to share in the unfolding of God's plan of creation brought into play those abilities and gifts which distinguish the human being from all other creatures. At the same time, their call established a fixed relationship between mankind and the rest of creation. Made in the image and likeness of God, Adam and Eve were to have exercised their dominion over the earth (Gen. 1:28) with wisdom and love. Instead, they destroyed the existing harmony *by deliberately going against the Creator's plan*, that is, by choosing to sin. This resulted not only in man's alienation from himself, in death and fratricide, but also in the earth's "rebellion" against him (cf. Gen. 3:17–19; 4:12). All of creation became subject to futility, waiting in a mysterious way to be set free and to obtain a glorious liberty together with all the children of God (cf. Rom. 8:20–21).

Christians believe that the Death and Resurrection of Christ accomplished the work of reconciling humanity to the Father, who "was pleased…through (Christ) to reconcile to himself *all things*, whether on earth or in heaven, making peace by the blood of his cross" (Col. 1:19–20). Creation was thus made new (cf. Rev. 21:5). Once subjected to the bondage of sin and decay (cf. Rom. 8:21), it has now received new life while "we wait for new heavens and a new earth in which righteousness dwells" (2 Pt. 3:13). Thus, the Father "has made known to us in all wisdom and insight the mystery…which he set forth in Christ as a plan for the fulness of time, to unite *all things* in him, all things in heaven and things on earth" (Eph. 1:9–10).

These biblical considerations help us to understand better *the relationship between human activity and the whole of creation*. When man turns his back on the Creator's plan, he provokes a disorder which has inevitable repercussions on the rest of the created order. If man is not at peace with God, then earth itself cannot be at peace: "Therefore the land mourns and all who dwell in it languish, and also the beasts of the field and the birds of the air and even the fish of the sea are taken away" (Hos. 4:3).

The profound sense that the earth is "suffering" is also shared by those who do not profess our faith in God. Indeed, the increasing devastation of the world of nature is apparent to all. It results from the behaviour of people who show a callous disregard for the hidden, yet perceivable requirements of the order and harmony which govern nature itself .

People are asking anxiously if it is still possible to remedy the damage which has been done. Clearly, an adequate solution cannot be found merely in a better management or a more rational use of the earth's resources, as important as these may be. Rather, we must go to the source of the problem and face in its entirety that profound moral crisis *of which the destruction of the environment is only one troubling aspect.*

II. The Ecological Crisis: A Moral Problem

Certain elements of today's ecological crisis reveal its moral character. First among these is the *indiscriminate* application of advances in science and technology. Many recent discoveries have brought undeniable benefits to humanity. Indeed, they demonstrate the nobility of the human vocation to participate *responsibly* in God's creative action in the world. Unfortunately, it is now clear that the application of these discoveries in the fields of industry and agriculture have produced harmful long-term effects. This has led to the painful realization that *we cannot interfere in one area of the ecosystem without paying due attention both to the consequences of such interference in other areas and to the well-being of future generations.*

The gradual depletion of the ozone layer and the related "greenhouse effect" has now reached crisis proportions as a consequence of industrial growth, massive urban concentrations and vastly increased energy needs. Industrial waste, the burning of fossil fuels, unrestricted deforestation, the use of certain types of herbicides, coolants and propellants: all of these are known to harm the atmosphere and environment. The resulting meteorological and atmospheric changes range from damage to health to the possible future submersion of low-lying lands.

While in some cases the damage already done may well be irreversible, in many other cases it can still be halted. It is necessary, however, that the entire human community—individuals, States and international bodies—take seriously the responsibility that is theirs.

The most profound and serious indication of the moral implications underlying the ecological problem is the lack of *respect for life* evident in many of the patterns of environmental pollution. Often, the interests of production prevail over concern for the dignity of workers, while economic interests take priority over the good of individuals and even entire peoples. In these cases, pollution or environmental destruction is the result of an unnatural and reductionist vision which at times leads to a genuine contempt for man.

On another level, delicate ecological balances are upset by the uncontrolled destruction of animal and plant life or by a reckless exploitation of natural resources. It should be pointed out that all of this, even if carried out in the name of progress and well-being, is ultimately to mankind's disadvantage.

Finally, we can only look with deep concern at the enormous possibilities of biological research. We are not yet in a position to assess the biological disturbance that could result from indiscriminate genetic manipulation and from the unscrupulous development of new forms of plant and animal life, to say nothing of unacceptable experimentation regarding the origins of human life itself. It is evident to all that in any area as delicate as this, indifference to fundamental ethical norms, or their rejection, would lead mankind to the very threshold of self-destruction.

Respect for life, and above all for the dignity of the human person, is the ultimate guiding norm for any sound economic, industrial or scientific progress.

The complexity of the ecological question is evident to all. There are, however, certain underlying principles, which, while respecting the legitimate autonomy and the specific competence of those involved, can direct research towards adequate and lasting solutions. These principles are essential to the building of a peaceful society; *no peaceful society can afford to neglect either respect for life or the fact that there is an integrity to creation.*

III. In Search of a Solution

Theology, philosophy and science all speak of a harmonious universe, of a "cosmos" endowed with its own integrity, its own internal, dynamic balance. *This order must be respected.* The human race is called to explore this order, to examine it with due care and to make use of it while safeguarding its integrity.

On the other hand, the earth is ultimately *a common heritage, the fruits of which are for the benefit of all.* In the words of the Second Vatican Council, "God destined the earth and all it contains for the use of every individual and all peoples" (*Gaudium et Spes*, 69). This has direct consequences for the problem at hand. It is manifestly unjust that a privileged few should continue to accumulate excess goods, squandering available resources, while masses of people are living in conditions of misery at the very lowest level of subsistence. Today, the dramatic threat of ecological breakdown is teaching us the extent to which greed and selfishness—both individual and collective—are contrary to the order of creation, an order which is characterized by mutual interdependence.

The concepts of an ordered universe and a common heritage both point to the necessity of a *more internationally coordinated approach to the management of the earth's goods.* In many cases the effects of ecological problems transcend the borders of individual States; hence their solution cannot be found solely on the national level. Recently there have been some promising steps towards such international action, yet the existing mechanisms and bodies are clearly not adequate for the development of a comprehensive plan of action. Political obstacles, forms of exaggerated nationalism and economic interests—to mention only a few factors—impede international cooperation and long-term effective action.

The need for joint action on the international level *does not lessen the responsibility of each individual State.* Not only should each State join with others in implementing internationally accepted standards, but it should also make or facilitate necessary socio-economic adjustments within its own borders, giving special attention to the most vulnerable sectors of society. The State should also actively endeavour within its own territory to prevent destruction of the atmosphere and biosphere, by carefully monitoring, among other things, the impact of new technological or scientific advances. The State also has the responsibility of ensuring that its citizens are not exposed to dangerous pollutants or toxic wastes. *The right to a safe environment* is ever more insistently presented today as a right that must be included in an updated Charter of Human Rights.

IV. The Urgent Need for a New Solidarity

The ecological crisis reveals the *urgent moral need for a new solidarity*, especially in relations between the developing nations and those that are highly industrialized. States must increasingly share responsibility, in complimentary ways, for the promotion of a natural and social environment that is both peaceful and healthy. The newly industrialized States cannot, for example, be asked to apply restrictive environmental standards to their emerging industries unless the industrialized States first apply them within their own boundaries. At the same time, countries in the process of industrialization are not morally free to repeat the errors made in the past by others, and recklessly continue to damage the environment through industrial pollutants, radical deforestation or unlimited exploitation of nonrenewable resources. In this context, there is urgent need to find a solution to the treatment and disposal of toxic wastes.

No plan or organization, however, will be able to effect the necessary changes unless world leaders are truly convinced of the absolute need for this new solidarity, which is demanded of them by the ecological crisis and which is essential for peace. *This need presents new opportunities for strengthening cooperative and peaceful relations among States.*

It must also be said that the proper ecological balance will not be found without *directly addressing the structural forms of poverty* that exist throughout the world. Rural poverty and unjust land distribution in many countries, for example, have led to subsistence farming and to the exhaustion of the soil. Once their land yields no more, many farmers move on to clear new land, thus accelerating uncontrolled deforestation, or they settle in urban centres which lack the infrastructure to receive them. Likewise, some heavily indebted countries are destroying their natural heritage, at the price of irreparable ecological imbalances, in order to develop new products for export. In the face of such situations it would be wrong to assign responsibility to the poor alone for the negative environmental consequences of their actions. Rather, the poor, to whom the earth is entrusted no less than to others, must be enabled to find a way out of their poverty. This will require a courageous reform of structures, as well as new ways of relating among peoples and States.

But there is another dangerous menace which threatens us, namely *war*. Unfortunately, modern science already has the capacity to change the environment for hostile purposes. Alterations of this kind over the long term could have unforeseeable and still more serious consequences. Despite the international agreements which prohibit chemical, bacteriological and biological warfare, the fact is that laboratory research continues to develop new offensive weapons capable of altering the balance of nature.

Today, any form of war on a global scale would lead to incalculable ecological damage. But even local or regional wars, however limited, not only destroy human life and social structures, but also damage the land, ruining crops and vegetation as well as poisoning the soil and water. The survivors of war are forced to begin a new life in very difficult environmental conditions, which in turn create situations of extreme social unrest, with further negative consequences for the environment.

Modern society will find no solution to the ecological problem unless it *takes a serious look at its lifestyle*. In many parts of the world society is given to instant gratification and consumerism while remaining indifferent to the damage which these cause. As I have already stated, the seriousness of the ecological issue lays bare the depth of man's moral crisis. If an appreciation of the value of the human person and of human life is lacking, we will also lose interest in others and in the earth itself. Simplicity, moderation and discipline, as well as a spirit of sacrifice, must become a part of everyday life, lest all suffer the negative consequences of the careless habits of a few.

An education in ecological responsibility is urgent: responsibility for oneself, for others, and for the earth. This education cannot be rooted in mere sentiment or empty wishes. Its purpose cannot be ideological or political. It must not be based on a rejection of the modern world or a vague desire to return to some "paradise lost." Instead, a true education in responsibility entails a genuine conversion in ways of thought and behaviour.

Churches and religious bodies, nongovernmental and governmental organizations, indeed all members of society, have a precise role to play in such education. The first educator, however, is the family, where the child learns to respect his neighbour and to love nature.

Finally, the aesthetic value of creation cannot be overlooked. Our very contact with nature has a deep restorative power; contemplation of its magnificence imparts peace and serenity. The Bible speaks again and again of the goodness and beauty of creation, which is called to glorify God (cf. Gen. 1:4ff; Ps. 8:2, 104:1ff; Wis. 13:3–5; Sir. 39:16, 33; 43:1, 9). More difficult perhaps, but no less profound, is the contemplation of the works of human ingenuity. Even cities can have a beauty all their own, one that ought to motivate people to care for their surroundings. Good urban planning is an important part of environmental protection, and respect for the natural contours of the land is an indispensable prerequisite for ecologically sound development. The relationship between a good aesthetic education and the maintenance of a healthy environment cannot be overlooked.

V. The Ecological Crisis: a Common Responsibility

Today the ecological crisis has assumed such proportions as to be *the responsibility of everyone*. As I have pointed out, its various aspects demonstrate the need for concerted efforts aimed at establishing the duties and obligations that belong to individuals, peoples, States and the international community. This not only goes hand in hand with efforts to build true peace, but also confirms and reinforces those efforts in a concrete way. When the ecological crisis is set within the broader context of *the search for peace* within society, we can understand better the importance of giving attention to what the earth and its atmosphere are telling us: namely, that there is an order in the universe which must be respected, and that the human person, endowed with the capability of choosing freely, has a grave responsibility to preserve this order for the well-being of future generations.

I wish to repeat that *the ecological crisis is a moral issue.*

Even men and women without any particular religious conviction, but with an acute sense of their responsibilities for the common good, recognize their obligation to contribute to the restoration of a healthy environment. All the more should men and women who believe in God the Creator, and who are thus convinced that there is a well-defined unity and order in the world, feel called to address the problem. Christians, in particular, realize that their responsibility within creation and their duty towards nature and the Creator are an essential part of their faith. As a result, they are conscious of a vast field of ecumenical and interreligious cooperation opening up before them.

At the conclusion of this Message, I should like to address directly my brothers and sisters in the Catholic Church, in order to remind them of their serious obligation to care for all of creation. The commitment of believers to a healthy environment for everyone stems directly from their belief in God the Creator, from their recognition of the effects of original and personal sin, and from the certainty of having been redeemed by Christ. Respect for life and for the dignity of the human person extends also to the rest of creation, which is called to join man in praising God (cf. Ps. 148:96).

In 1979, I proclaimed St. Francis of Assisi as the heavenly Patron of those who promote ecology (cf. Apostolic Letter *Inter Sanctos: AAS* 71 [1979], 1509f.). He offers Christians an example of genuine and deep respect for the integrity of creation. As a friend of the poor who was loved by God's creatures, St. Francis invited all of creation—animals, plants, natural forces, even Brother Sun and Sister Moon—to give honour and praise to the Lord. The poor man of Assisi gives us striking witness that when we are at peace with God we are better able to devote ourselves to building up that peace with all creation which is inseparable from peace among all peoples.

It is my hope that the inspiration of St. Francis will help us to keep ever alive a sense of "fraternity" with all those good and beautiful things which Almighty God has created. And may he remind us of our serious obligation to respect and watch over them with care, in light of that greater and higher fraternity that exists within the human family.

From the Vatican, 8 December 1989.

Common Declaration on Environmental Ethics of Pope John Paul II and the Ecumenical Patriarch His Holiness Bartholomew I

Monday, June 10, 2002

We are gathered here today in the spirit of peace for the good of all human beings and for the care of creation. At this moment in history, at the beginning of the third millennium, we are saddened to see the daily suffering of a great number of people from violence, starvation, poverty and disease. We are also concerned about the negative consequences for humanity and for all creation resulting from the degradation of some basic natural resources such as water, air and land, brought about by an economic and technological progress which does not recognize and take into account its limits.

Almighty God envisioned a world of beauty and harmony, and He created it, making every part an expression of His freedom, wisdom and love (cf. Gen. 1:1–25).

At the centre of the whole of creation, He placed us, human beings, with our inalienable human dignity. Although we share many features with the rest of the living beings, Almighty God went further with us and gave us an immortal soul, the source of self-awareness and freedom, endowments that make us in His image and likeness (cf. Gen. 1:26–31;2:7). Marked with that resemblance, we have been placed by God in the world in order to cooperate with Him in realizing more and more fully the divine purpose for creation.

At the beginning of history, man and woman sinned by disobeying God and rejecting His design for creation. Among the results of this first sin was the destruction of the original harmony of creation. If we examine carefully the social and environmental crisis which the world community is facing, we must conclude that we are still betraying the mandate God has given us: to be stewards called to collaborate with God in watching over creation in holiness and wisdom.

God has not abandoned the world. It is His will that His design and our hope for it will be realized through our cooperation in restoring its original harmony. In our own time we are witnessing a growth of an *ecological awareness* which needs to be encouraged, so that it will lead to practical programmes and initiatives.

An awareness of the relationship between God and humankind brings a fuller sense of the importance of the relationship between human beings and the natural environment, which is God's creation and which God entrusted to us to guard with wisdom and love (cf. Gen. 1:28).

Respect for creation stems from respect for human life and dignity. It is on the basis of our recognition that the world is created by God that we can discern an objective moral order within which to articulate a code of environmental ethics. In this perspective, Christians and all other believers have a specific role to play in proclaiming moral values and in educating people in *ecological awareness*, which is none other than responsibility towards self, towards others, towards creation.

What is required is an act of repentance on our part and a renewed attempt to view ourselves, one another, and the world around us within the perspective of the divine design for creation. The problem is not simply economic and technological; it is moral and spiritual. A solution at the economic and technological level can be found only if we undergo, in the most radical way, an inner change of heart, which can lead to a change in lifestyle and of unsustainable patterns of consumption and production. A genuine *conversion* in Christ will enable us to change the way we think and act.

First, we must regain humility and recognize the limits of our powers, and most importantly, the limits of our knowledge and judgment. We have been making decisions, taking actions and assigning values that are leading us away from the world as it should be, away from the design of God for creation, away from all that is essential for a healthy planet and a healthy commonwealth of people.

A new approach and a new culture are needed, based on the centrality of the human person within creation and inspired by environmentally ethical behavior stemming from our triple relationship to God, to self and to creation. Such an ethics fosters interdependence and stresses the principles of universal solidarity, social justice and responsibility, in order to promote a true culture of life.

Secondly, we must frankly admit that humankind is entitled to something better than what we see around us. We and, much more, our children and future generations are entitled to a better world, a world free from degradation, violence and bloodshed, a world of generosity and love.

Thirdly, aware of the value of prayer, we must implore God the Creator to enlighten people everywhere regarding the duty to respect and carefully guard creation.

We therefore invite all men and women of good will to ponder the importance of the following ethical goals:

1. To think of the world's children when we reflect on and evaluate our options for action.
2. To be open to study the true values based on the natural law that sustain every human culture.
3. To use science and technology in a full and constructive way, while recognizing that the findings of science have always to be evaluated in the light of the centrality of the human person, of the common good

and of the inner purpose of creation. Science may help us to correct the mistakes of the past, in order to enhance the spiritual and material well-being of the present and future generations.

It is love for our children that will show us the path that we must follow into the future.

4. To be humble regarding the idea of ownership and to be open to the demands of solidarity. Our mortality and our weakness of judgement together warn us not to take irreversible actions with what we choose to regard as our property during our brief stay on this earth. We have not been entrusted with unlimited power over creation, we are only stewards of the common heritage.

5. To acknowledge the diversity of situations and responsibilities in the work for a better world environment. We do not expect every person and every institution to assume the same burden. Everyone has a part to play, but for the demands of justice and charity to be respected the most affluent societies must carry the greater burden, and from them is demanded a sacrifice greater than can be offered by the poor.

Religions, governments and institutions are faced by many different situations; but on the basis of the principle of subsidiarity all of them can take on some tasks, some part of the shared effort.

6. To promote a peaceful approach to disagreement about how to live on this earth, about how to share it and use it, about what to change and what to leave unchanged. It is not our desire to evade controversy about the environment, for we trust in the capacity of human reason and the path of dialogue to reach agreement.

We commit ourselves to respect the views of all who disagree with us, seeking solutions through open exchange, without resorting to oppression and domination.

It is not too late. God's world has incredible healing powers. Within a single generation, we could steer the earth toward our children's future. Let that generation start now, with God's help and blessing.

Address to Young Muslims

Morocco
August 19, 1985

This address is emblematic of Pope John Paul II's speeches to all youth and to his spirit of interfaith dialogue. This was not his only address to Muslim youth or those of other faiths. He addressed the young Muslims of Casablanca and other places, attended interfaith gatherings with Muslims, Hindus, and Christians in the Sahel, and held audiences with people of varied faiths from around the world. In this address, you can hear how he seeks the commonality of a shared Creator and earth, shared hopes and dreams for humanity, and for a better, more just and harmonious world for all: his clear ecological vision.

Dear Young People,

I give thanks and glory to God who has granted that I should meet with you today. His Majesty the King did me the honour of visiting me in Rome some years ago, and he had the courtesy to invite me to visit your country and meet you. I joyfully accepted the invitation from the Sovereign of this country to speak with you in this Year of Youth.

I often meet young people, usually Catholics. It is the first time that I find myself with young Muslims.

Christians and Muslims, we have many things in common, as believers and as human beings. We live in the same world, marked by many signs of hope, but also by multiple signs of anguish. For us, Abraham is a very model of faith in God, of submission to his will and of confidence in his goodness. We believe in the sane God, the one God, the living God, the God who created the world and brings his creatures to their perfection.

It is therefore towards this God that my thought goes and that my heart rises: it is of God himself that, above all, I wish to speak with you; of him, because it is in him that we believe, you Muslims and we Catholics. I wish also to speak with you about human values, which have their basis in God, these values which concern the blossoming of our person, as also that of our families and our societies, as well as that of the international community. The mystery of God, is it not the highest reality from which depends the very meaning which man gives to his life? And is it not the first problem that presents itself to a young person, when he reflects upon the mystery of his own existence and on the values which he intends to choose in order to build his growing personality?

For my part, in the Catholic Church, I bear the responsibility of the successor of Peter, the Apostle chosen by Jesus to strengthen his brothers in the faith. Following the Popes who succeeded one another uninterruptedly in the passage of history, I am today the Bishop of Rome, called to be, among his brethren in the world, the witness of the Christian faith and the guarantee of the unity of all the members of the Church.

Also, it is as a believer that I come to you today. It is quite simply that I would like to give here today the witness of that which I believe, of that which I wish for the well-being of the people, my brothers, and of the people, my brothers, and of that which, from experience, I consider to be useful for all.

First of all, I invoke the Most High, the all-powerful God who is our creator. He is the origin of all life, as he is at the source of all that is good, of all that is beautiful, of all that is holy.

He separated the light from the darkness. He caused the whole universe to grow in a marvellous order. He willed that the plants should grow and bear fruit, just as he willed that the birds of the sky, the animals of the earth and the fish of the sea should multiply.

He made us, us men, and we are from him. His holy law guides our life. It is the light of God which orientates our destiny and enlightens our conscience. He renders us capable of loving and of transmitting life. He asks every man to respect every human creature and to love him as a friend, a companion, a brother. He invites us to help him when he is wounded, when he is abandoned, when he is hungry and thirsty, in short, when he no longer knows where to find his direction on the pathways of life.

Yes, God asks that we should listen to his voice. He expects from us obedience to his holy will in a free consent of mind and of heart.

That is why we are accountable before him. It is he, God, who is our judge; he who alone is truly just. We know, however, that his mercy is inseparable from his justice. When man returns to him, repentant and contrite, after having strayed away into the disorder of sin and the works of death, God then reveals himself as the One who pardons and shows mercy.

To him, therefore, our love and our adoration! For his blessing and for his mercy, we thank him, at all times and in all places.

In a world which desires unity and peace, and which however experiences a thousand tensions and conflicts, should not believers favour friendship between the men and the peoples who form one single community on earth? We know that they have one and the same origin and one and the same final end: the God who made them and who waits for them, because he will gather them together.

For its part, the Catholic Church, twenty years ago at the time of the Second Vatican Council, undertook in the person of its bishops, that is, of its religious leaders, to seek collaboration between the believers. It published a *document on dialogue between the religions* (Nostra Aetate). It affirms that all men, especially those of living faith, should respect each other, should rise above all discrimination, should live in harmony and serve the universal brotherhood (cf. document cited above, n. 5). The

Church shows particular attention to the believing Muslims, given their faith in the one God, their sense of prayer, and their esteem for the moral life (cf. n. 3). It desires that Christians and Muslims together "promote harmony for all men, social justice, moral values, peace, liberty" (ibid.).

Dialogue between Christians and Muslims is today more necessary than ever. It flows from our fidelity to God and supposes that we know how to recognize God by faith, and to witness to him by word and deed in a world ever more secularized and at times even atheistic.

The young can build a better future if they first put their faith in God and if they pledge themselves to build this new world in accordance with God's plan, with wisdom and trust.

Today we should *witness* to the spiritual values of which the world has need. The first is *our faith in God*.

God is the source of all joy. We should also witness to our worship of God, by our adoration, our prayer of praise and supplication. Man cannot live without prayer, any more than he can live without breathing. We should witness to our *humble search for his will*; it is he who should inspire our pledge for a more just and more united world. God's ways are not always our ways. They transcend our actions, which are always incomplete, and the intentions of our heart, which are always imperfect. God can never be used for our purposes, for he is above all.

This witness of faith, which is vital for us and which can never tolerate either infidelity to God or indifference to the truth, is made with respect for the other religious traditions, because everyone hopes to be respected for what he is in fact, and for what he conscientiously believes. We desire that all may reach the fullness of the divine truth, but no one can do that except through the free adherence of conscience, protected from exterior compulsions which would be unworthy of the free homage of reason and of heart which is characteristic of human dignity. There, is the true meaning of religious liberty, which at the same time respects God and man. It is the sincere veneration of such worshippers that God awaits, of worshippers in spirit and in truth.

We are convinced that "we cannot truly pray to God the Father of all mankind, if we treat any people in other than brotherly fashion, for all mankind is created in God's image" (*Nostra Aetate*, n. 5).

Therefore we must also *respect, love and help every human being*, because he is a creature of God and, in a certain sense, his image and his representative, because he is the road leading to God, and because he does not fully fulfill himself unless he knows God, unless he accepts him with all his heart, and unless he obeys him to the extent of the ways of perfection.

Furthermore, this obedience to God and this love for man should lead us *to respect man's rights*, these rights which are the expression of God's will and the demands of human nature such as it was created by God.

Therefore, respect and dialogue require reciprocity in all spheres, especially in that which concerns basic freedoms, more particularly religious freedom. They favour peace and agreement between the peoples. They help to resolve together the problems of today's men and women, especially those of the young.

Normally the young look towards the future, they long for a more just and more human world. God made young people such, precisely that they might help to transform the world in accordance with his plan of life. But to them, too, the situation often appears to have its shadows.

In this world there are frontiers and divisions between men, as also misunderstandings between the generations; there are, likewise, racism, wars and injustices, as also hunger, waste and unemployment. These are the dramatic evils which touch us all, more particularly the young of the entire world. Some are in danger of discouragement, others of capitulation, others of willing to change everything by violence or by extreme solutions. Wisdom teaches us that self-discipline and love are then the only means to the desired renewal.

God does not will that people should remain passive. He entrusted the earth to them that together they should subdue it, cultivate it, and cause it to bear fruit.

You are charged with the world of tomorrow. It is *by fully and courageously undertaking your responsibilities* that you will be able to overcome the existing difficulties. It reverts to you to take the initiatives and not to wait for everything to come from the older people and from those in office. You must build the world and not just dream about it.

It is *by working in harmony* that one can be effective. Work properly understood is a service to others. It creates links of solidarity. The experience of working in common enables one to purify oneself and to discover the richness of others. It is thus that, gradually, a climate of trust can be born which enables each one to grow, to expand, and "to be more." Do not fail, dear young people, to collaborate *with the adults*, especially with your parents and teachers as well as with the "leaders" of society and of the State. The young should not isolate themselves from the others. The young need the adults, just as the adults need the young.

In this working together, the human person, man or woman, should never be sacrificed. *Each person is unique* in God's eyes. Each one ought to be appreciated for what he is, and, consequently, respected as such. No one should make use of his fellow man; no one should exploit his equal; no one should contemn his brother.

It is in these conditions that a more human, more just, and more fraternal world will be able to be born, a world where each one can find his place in dignity and freedom. It is this world of the twenty-first century that is in your hands; it will be what you make it.

This world, which is about to come, depends on the *young people of all the countries of the world.* Our world is divided, and even shattered; it experiences multiple conflicts and grave injustices. There is no real North-South solidarity; there is not enough mutual assistance between the nations of the South. There are in the world cultures and races which are not respected.

Why is all this? *It is because people do not accept their differences*: they do not know each other sufficiently. They reject those who have not the same civilization. They refuse to help each other. They are unable to free themselves from egoism and from self-conceit.

But God created all men equal in dignity, though different with regard to gifts and to talents. Mankind is a whole where each one has his part to play; the worth of the various peoples and of the diverse cultures must be recognized. The world is as it were a living organism; each one has something to receive from the others, and has something to give to them.

I am happy to meet you here in Morocco. Morocco has a *tradition of openness*. Your scholars have travelled, and you have welcomed scholars from other countries. Morocco has been a meeting place of civilizations: it has permitted exchanges with the East, with Spain, and with Africa. Morocco has a *tradition of tolerance*; in this Muslim country there have always been Jews and nearly always Christians; that tradition has been carried out in respect, in a positive manner. You have been, and you remain, a hospitable country. You, young Moroccans, are then prepared to become citizens of tomorrow's world, of this *fraternal world* to which, with the young people of all the world, you aspire.

I am sure that all of you, young people, are capable of this dialogue. You do not wish to be conditioned by prejudices. You are ready to build a civilization based on love. You can work to cause the barriers to fall, barriers that are due at times to pride, but more often to man's feebleness and fear. You wish to love others, without any limit of nation, race or religion.

For that, *you want justice and peace*. "Peace and youth go forward together," as I said in my message for this year's World Day of Peace. You do not want either war or violence. You know the price that they cause innocent people to pay. Neither do you want the escalation of armaments. That does not mean that you wish to have peace at any price. Peace goes side by side with justice. You do not want anyone to be oppressed. You want peace in justice.

First of all, you wish that people should have enough on which to live. Young people who have the good fortune to pursue their studies have the right to be solicitous about the profession that they will be able to exercise on their your own behalf. But they also must concern themselves with the living conditions, often more difficult, of their brothers and sisters who live in the same country, and indeed in the whole world. How can one remain indifferent, in fact, when other human beings, in great numbers, die of hunger, of malnutrition or lack of health help, when they suffer cruelly from drought, when they are reduced to unemployment or to emigration through economic laws that are beyond their control, when they endure the precarious situation of refugees, packed into camps, as a consequence of human conflicts? God has given the earth to mankind as a whole in order that people might jointly draw their subsistence from it, and that every people might have the means to nourish itself, to take care of itself; and to live in peace.

But important as the economic problems may be, man does not live on bread alone, he needs an intellectual and spiritual life; it is there that he finds the soul of this new world to which you aspire. Man has need to develop *his spirit and his conscience*. This is often lacking to the man of today. Forgetfulness of values and the crisis of identity which frustrate our world oblige us to excel ourselves in a renewed effort of research and investigation. The interior light which will thus be born in our conscience will enable meaning to be given to development, to orientate it towards the good of man, of every man and of all men, in accordance with God's plan.

The Arabs of the Mashriq and the Maghrib, and Muslims in general, have a long tradition of study and of erudition: literary, scientific, philosophic. You are the heirs to this tradition, you must study in order to learn to know this world which God has given us, to understand it, to discover its meaning, with a desire and a *respect for truth*, and in order to learn to know the peoples and the men created and loved by God, so as to prepare yourselves better to serve them.

Still more, the search for truth will lead you, beyond intellectual values, to the spiritual dimension of the interior life.

Man is a *spiritual being*. We, believers, know that we do not live in a closed world. We believe in God. We are worshippers of God. We are seekers of God.

The Catholic Church regards with respect and *recognizes the quality of your religious progress*, the richness of your spiritual tradition.

We Christians, also, are proud of our own religious tradition.

I believe that we, Christians and Muslims, must recognize with joy the religious values that we have in common, and give thanks to God for them. Both of us believe in one God the only God, who is all Justice and all Mercy; we believe in the importance of prayer, of fasting, of almsgiving, of repentance and of pardon; we believe that God will be a merciful judge to us at the end of time, and we hope that after the resurrection he will be satisfied with us and we know that we will be satisfied with him.

Loyalty demands also that we should recognize and respect our differences. Obviously the most fundamental is the view that we hold on the person and work of Jesus of Nazareth. You know that, for the Christians, this Jesus causes them to enter into an intimate knowledge of the mystery of God and into a filial communion by his gifts, so that they recognize him and proclaim him Lord and Saviour.

Those are important differences, which we can accept with humility and respect, in mutual tolerance; there is a mystery there on which, I am certain, God will one day enlighten us.

Christians and Muslims, in general we have badly understood each other, and sometimes, in the past, we have opposed and even exhausted each other in polemics and in wars.

I believe that, today, God invites us *to change our old practices.* We must respect each other, and also we must stimulate each other in good works on the path of God.

With me, you know what is the reward of spiritual values. Ideologies and slogans cannot satisfy you nor can they solve the problems of your life. Only the spiritual and moral values can do it, and they have God as their fundament.

Dear young people, I wish that you may be able to help in thus building a world where God may have first place in order to aid and to save mankind. On this path, you are assured of the esteem and the collaboration of your Catholic brothers and sisters whom I represent among you this evening.

I should now like to thank His Majesty the King for having invited me. I thank you also, dear young people of Morocco, for having come here and listened with confidence to my witness.

But still more, I would like to thank God who permitted this meeting. We are all in his sight. Today he is the first witness of our meeting. It is he who puts in our hearts the feelings of mercy and understanding, of pardon and of reconciliation, of service and of collaboration. Must not the believers that we are reproduce in their life and in their city the Most Beautiful Names which our religious traditions recognize for him? May we then be able to be available for him, and to be submissive to his will, to the calls that he makes to us! In this way our lives will find a new dynamism.

Then, I am convinced, a world can be born where men and women of living and effective faith will sing to the glory of God, and will seek to build a human society in accordance with God's will.

I should like to finish by invoking him personally in your presence:

O God, you are our creator.
You are limitlessly good and merciful.
To You is due the praise of every creature.
O God, You have given to us an interior law by which we should live.
To do Your will is to perform our task.
To follow Your ways is to find peace of soul.
To You we offer our obedience.
Guide us in all the steps that we undertake on earth.
Free us from evil inclinations which turn our heart from Your will.
Do not permit that in invoking Your Name we should ever justify the human
   disorders.

O God, you are the One Alone to whom we make our adoration.

Do not permit that we should estrange ourselves from You.

O God, judge of all mankind, help us to belong to Your elect on the last day.

O God, author of justice and peace, grant us true joy and authentic love, as also a lasting fraternity among all peoples.

Fill us with Your gifts for ever. Amen!

Address to Aborigines and Torres Strait Islanders

Blatherskite Park, Alice Springs (Australia)
November 29, 1986

This address represents similar themes that Pope John Paul II raised with native peoples the world over, as their struggles are so similar. It illustrates the power, poetry, and emotion with which he spoke about the injustices that have been and are still being perpetrated. It also highlights what all people should learn from the values and ways of the traditional aboriginal and indigenous peoples of the world, who live as kin with the land and water, animals and plants—all of God's creation.

Dear Brothers and Sisters,

It is a great joy for me to be here today in Alice Springs and to meet so many of you, the Aborigines and Torres Strait Islanders of Australia. I want to tell you right away how much the Church esteems and loves you, and how much she wishes to assist you in your spiritual and material needs.

At the beginning of time, as God's Spirit moved over the waters, he began to communicate something of his goodness and beauty to all creation. When God then created man and woman, he gave them the good things of the earth for their use and benefit; and he put into their hearts abilities and powers, which were his gifts. And to all human beings throughout the ages God has given a desire for himself, a desire which different cultures have tried to express in their own ways.

As the human family spread over the face of the earth, your people settled and lived in this big country that stood apart from all the others. Other people did not even know this land was here; they only knew that somewhere in the southern oceans of the world there was "The Great South Land of the Holy Spirit."

But for thousands of years you have lived in this land and fashioned a culture that endures to this day. And during all this time, the Spirit of God has been with you. Your "Dreaming," which influences your lives so strongly that, no matter what happens, you remain for ever people of your culture, is your only way of touching the mystery of God's Spirit in you and in creation. You must keep your striving for God and hold on to it in your lives.

The rock paintings and the discovered evidence of your ancient tools and implements indicate the presence of your age-old culture and prove your ancient occupancy of this land.

Your culture, which shows the lasting genius and dignity of your race, must not be allowed to disappear. Do not think that your gifts are worth so little that you should no longer bother to maintain them. Share them with each other and teach them to your children. Your songs, your stories, your paintings, your dances, your languages, must never be lost. Do you perhaps remember those words that Paul VI spoke to the aboriginal people during his visit to them in 1970? On that occasion he said: "We know that you have a life style proper to your own ethnic genius or culture—a culture which the Church respects and which she does not in any way ask you to renounce...Society itself is enriched by the presence of different cultural and ethnic elements. For us you and the values you represent are precious. We deeply respect your dignity and reiterate our deep affection for you."

For thousands of years this culture of yours was free to grow without interference by people from other places. You lived your lives in spiritual closeness to the land, with its animals, birds, fishes, waterholes, rivers, hills and mountains. Through your closeness to the land you touched the sacredness of man's relationship with God, for the land was the proof of a power in life greater than yourselves.

You did not spoil the land, use it up, exhaust it. And then walk away from it. You realized that your land was related to the source of life.

The silence of the Bush taught you a quietness of soul that put you in touch with another world, the world of God's Spirit. Your careful attention to the details of kinship spoke of your reverence for birth, life and human generation. You knew that children need to be loved, to be full of joy. They need a time to grow in laughter and to play, secure in the knowledge that they belong to their people.

You had a great respect for the need which people have for law, as a guide to living fairly with each other. So you created a legal system—very strict it is true—but closely adapted to the country in which you lived your lives. It made your society orderly. It was one of the reasons why you survived in this land.

You marked the growth of your young men and women with ceremonies of discipline that taught them responsibility as they came to maturity.

These achievements are indications of human strivings. And in these strivings you showed a dignity open to the message of God's revealed wisdom to all men and women, which is the great truth of the Gospel of Jesus Christ.

Some of the stories from your Dreamtime legends speak powerfully of the great mysteries of human life, its frailty, its need for help, its closeness to spiritual powers and the value of the human person. They are not unlike some of the great inspired lessons from the people among whom Jesus himself was born. It is wonderful to see how people, as they accept the Gospel of Jesus, find points of agreement between their own traditions and those of Jesus and his people.

The culture which this long and careful growth produced was not prepared for the sudden meeting with another people, with different customs and traditions, who came to your country nearly 200 years ago. They were different from Aboriginal people. Their traditions, the organization of their lives, and their attitudes to the land were quite strange to you. Their law too was quite different. These people had knowledge, money and power; and they brought with them some patterns of behaviour from which the Aboriginal people were unable to protect themselves.

The effects of some of those forces are still active among you today. Many of you have been dispossessed of your traditional lands, and separated from your tribal ways, though some of you still have your traditional culture. Some of you are establishing Aboriginal communities in the towns and cities. For others there is still no real place for campfires and kinship observances except on the fringes of country towns. There, work is hard to find, and education in a different cultural background is difficult. The discrimination caused by racism is a daily experience.

You have learned how to survive, whether on your own lands, or scattered among the towns and cities. Though your difficulties are not yet over, you must learn to draw on the endurance which your ancient ceremonies have taught you. Endurance brings with it patience; patience helps you to find the way ahead, and gives you courage for your journey.

Take heart from the fact that many of your languages are still spoken and that you still possess your ancient culture. You have kept your sense of brotherhood. If you stay closely united, you are like a tree standing in the middle of a bush-fire sweeping through the timber. The leaves are scorched and the tough bark is scarred and burned; but inside the tree the sap is still flowing, and under the ground the roots are still strong. Like that tree you have endured the flames, and you still have the power to be reborn. The time for this rebirth is now!

We know that during the last two hundred years certain people tried to understand you, to learn about you, to respect your ways and to honour you as persons. These men and women, as you soon realized, were different from others of their race. They loved and cared for the indigenous people. They began to share with you their stories of God, helped you cope with sickness, tried to protect you from ill-treatment. They were honest with you, and showed you by their lives how they tried to avoid the bad things in their own culture. These people were not always successful, and there were times when they did not fully understand you. But they showed you good will and friendship. They came from many different walks of life. Some were teachers and doctors and other professional people; some were simple folk. History will remember the good example of their charity and fraternal solidarity.

Among those who have loved and cared for the indigenous people, we especially recall with profound gratitude all the missionaries of the Christian faith. With immense generosity they gave their lives in service to you and to your forebears. They helped to educate the Aboriginal people and offered health and social services. Whatever their human frailty, and whatever mistakes they may have made, nothing can ever minimize the depth of their charity. Nothing can ever cancel out their greatest contribution, which was to proclaim to you Jesus Christ and to establish his Church in your midst.

From the earliest times men like Archbishop Polding of Sydney opposed the legal fiction adopted by European settlers that this land was *terra nullius*—nobody's country. He strongly pleaded for the rights of the Aboriginal inhabitants to keep the traditional lands on which their whole society depended. The Church still supports you today.

Let it not be said that the fair and equitable recognition of Aboriginal rights to land is discrimination. To call for the acknowledgment of the land rights of people who have never surrendered those rights is not discrimination. Certainly, what has been done cannot be undone. But what can now be done to remedy the deeds of yesterday must not be put off till tomorrow.

Christian people of good will are saddened to realize—many of them only recently—for how long a time Aboriginal people were transported from their homelands into small areas or reserves where families were broken up, tribes split apart, children orphaned and people forced to live like exiles in a foreign country.

The reserves still exist today, and require a just and proper settlement that still lies unachieved. The urban problems resulting from the transportation and separation of people still have to be addressed, so that these people may make a new start in life with each other once again.

The establishment of a new society for Aboriginal people cannot go forward without just and mutually recognized agreements with regard to these human problems, even though their causes lie in the past. The greatest value to be achieved by such agreements, which must be implemented without causing new injustices, is respect for the dignity and growth of the human person. And you, the Aboriginal people of this country and its cities, must show that you are actively working for your own dignity of life. On your part, you must show that you too can walk tall and command the respect which every human being expects to receive from the rest of the human family.

The Gospel of our Lord Jesus Christ speaks all languages. It esteems and embraces all cultures. It supports them in everything human and, when necessary, it purifies them. Always and everywhere the Gospel uplifts and enriches cultures with the revealed message of a loving and merciful God.

That Gospel now invites you to become, through and through, Aboriginal Christians. It meets your deepest desires. You do not have to be people divided into two parts, as though an Aboriginal had to borrow the faith and life of Christianity, like a hat or a pair of shoes, from someone else who owns them. Jesus calls you to accept his words and his values into your own culture. To develop in this way will make you more than ever truly Aboriginal.

The old ways can draw new life and strength from the Gospel. The message of Jesus Christ can lift up your lives to new heights, reinforce all your positive values and add many others, which only the Gospel in its originality proposes. Take this

Gospel into your own language and way of speaking; let its spirit penetrate your communities and determine your behaviour towards each other, let it bring new strength to your stories and your ceremonies. Let the Gospel come into your hearts and renew your personal lives.

The Church invites you to express the living word of Jesus in ways that speak to your Aboriginal minds and hearts. All over the world people worship God and read his word in their own language, and colour the great signs and symbols of religion with touches of their own traditions. Why should you be different from them in this regard, why should you not be allowed the happiness of being with God and each other in Aboriginal fashion?

As you listen to the Gospel of our Lord Jesus Christ, seek out the best things of your traditional ways. If you do, you will come to realize more and more your great human and Christian dignity. Let your minds and hearts be strengthened to begin a new life now. Past hurts cannot be healed by violence, nor are present injustices removed by resentment. Your Christian faith calls you to become the best kind of Aboriginal people you can be. This is possible only if reconciliation and forgiveness are part of your lives. Only then will you find happiness. Only then will you make your best contribution to all your brothers and sisters in this great nation. You are part of Australia and Australia is part of you. And the Church herself in Australia will not be fully the Church that Jesus wants her to be until you have made your contribution to her life and until that contribution has been joyfully received by others.

In the new world that is emerging for you, you are being called to live fully human and Christian lives, not to die of shame and sorrow. But you know that to fulfil your role you need a new heart. You will already feel courage rise up inside you when you listen to God speaking to you in these words of the Prophets:

"Do not be afraid for I have redeemed you; I have called you by your name, you are mine. Do not be afraid, for I am with you.".

And again:

"I am going to...gather you together...and bring you home to your own land...I shall give you a new heart and put a new spirit in you...You shall be my people and I will be your God."

With you I rejoice in the hope of God's gift of salvation, which has its beginnings here and now, and which also depends on how we behave towards each other, on what we put up with, on what we do, on how we honour God and love all people.

Dear Aboriginal people: the hour has come for you to take on new courage and new hope. You are called to remember the past, to be faithful to your worthy traditions, and to adapt your living culture whenever this is required by your own needs and those of your fellowman. Above all you are called to open your hearts ever more to the consoling, purifying and uplifting message of Jesus Christ, the Son of God, who died so that we might all have life, and have it to the full.

Acknowledgments

Over 12 years, there are so many to thank.

Without the ongoing perseverance, idealism, and supportive friendship of my agent, Edythea Ginis Selman, this book would never have seen the light of day. And if my editor, Alessandra Lusardi, had not had the vision to take a risk and use her insightful professional skills to champion and guide this project, again, it could not have happened. Your timing was exquisite and you were worth waiting for! Thanks to all at Rizzoli. Bill McKibben, you are the gracious, tenacious and kind Martin Luther King, Jr. of climate rights, and your sacrificial devotion to alerting the world of the realities and effects of climate change gives us all hope. Thanks so much for organizing the 350. org movement to give us ways to get to work creatively and effectively!

To Frederick W. Kreuger, from the National Religious Coalition on Creation Care (NRCCC) who guided me to this project, even when I was unwilling. You have been its guardian angel. To Sister Marjorie Keenan, RSHM, whose pioneering work to gather together the ecological statements from Pope John Paul II and the Holy See formed the initial core of this project. And to Curt Meine, who led me to Aldo Leopold and to Fred, and to good friends. Like McKibben, your dedication to passing on a Leopoldian ethic and restored world, with the relationships between people and nature reestablished, is beyond compare, and your way of music-making and networking others for friendship and action make the work sweet. Thank you for your friendship and assistance.

To Tebaldo Vinciguerra, from the Pontifical Council of Peace and Justice, who tracked me down and gave me the nudge I needed to keep going, and who became a friend; To His Excellency Cardinal Peter Turkson and Most Reverend Bishop William S. Skylstad, who gave me such kind encouragement; to my dear Allen Johnson, from Christians for the Mountains; Patrick Carolan, from the Franciscan Action Network; and David Krantz, of the Green Zionist Alliance. You are each amazing, and your kindness, mentoring, and friendship have meant so much.

Yet how could I have done this long project without the ongoing support of my treasured friends: Julie Dunlap and Lucy Perez, Mary Beth Nierengarten, Pamela Hall, Nadia Higgins, Denise Herrmann, Gail Lamberty, Jules Hermes, and my dear, dear, dear Peeps? Claire Zajac, Kiki Gorbatenko Roth, and Mary Pat Finnegan, with research, technical, and friendship assistance from Brian Huffman, Charlie Roth, and Michael Rosenfeld. How many times did I bug you all to read things or carry me through despair? Your devotion to me and this project was beyond compare! (And Claire, you live up to your namesake. Your determination to get this book published and carry me through was beyond belief.) And to my awesome book club friends without whom life would be so dull: Mary Prindiville, Ellen Hanten, Joan Bushman, Tracy Habisch-Ahlin, Chris Cameron, Carol Hardin, Brenda Bredahl, and Michelle....

And what of my long-time mentors Sisters Brigid and Jane McDonald (CSJ) and harmonica-playing Aunt Mary Williams who taught me so much about living?

Thank you to long-time friends Paul Rome and Leigh Beatty, for always generously giving David and me soul-expanding time at your cabin in Ely, gateway to the Boundary Waters and Quetico Wilderness Areas. It's always a taste of heaven. And to all of you from the Brew Crew. How fortunate I was to marry into your group!

Thank you for your inspiration Sister Johannes Klaus, Father Ben Colucci, the late Sisters Bernadette Kalscheur and Helen Mrosla (all Franciscans!), Father John Parr, Father Patrick McConnell, Father Gene Murphy, Father Gerald P. Harris, Father John Gerlitz, the very loving, faith-generating Carmelites of Hudson, Father William Poole, Felix Lopez, Sister Anne Thompson (OSB), Molly Druffner, Father Mansuetus Setonga, Friends of Mater Dei, Lake Atitlan Libraries, and the San Jose de Tesero de Yalpamech Sister Parish team. Thanks also to Ruth Haag Brombach and Chris Klejbuk from St. Catherine University Alumnae Association.

To Paul Hawken; the late, great Wangari Maathai; Polar Explorer Will Steger; Sarah Susanka; Mary Doria Russell from far away, and closer to home, Mary Seyer, Mary Piasecki, Larry Huiras, and Marshal Toman—your kind words at key low moments (I had so many!) kept me going and were worth more than can be noted.

And my gratitude to copyeditors Michael Burke and Jane Cavolina, who helped clean up this complicated manuscript, and the persevering type-setter and cover designer Sara Stemen, who found that perfect photograph of Pope John Paul II—what magic!

Finally, and primarily, *absolutely not least*, to my incredible family: my parents Marilyn and Rudy Lorbiecki, my siblings Mark, Kate, Jean, John and LeAnn, and your terrific spouses (Yoshimi, Rick, Patty), Dewey Nelson and Chris Kalmbach, Jim and Patty Carlen (Peter, Sam, and Will), Aunt Dorothy and Bob Berg, and my fabulous nieces and nephews—Alex, Abby, Lawrence, Bella, Rachel (and Jeremy), Sarah (April, Sophia and Jeff), Grace, Emily, Max, Peter, Steven, Claire, Samantha, and Will. You have so cheered me on at every step of the way. And to my wonderful in-laws and more nieces and nephews: Pete Mataya and Mary Ann Ulishney and Myrna Olsen, and all the wonderful Matayas—Carol, Dan, Lisa, Bruce, Abby, Emma, Liam, Darryl, Rebecca, Becca, Kevin, Myles and Silas, Michael (and Abby in the wings), Safaie and Kerianna. And all my so-beloved Lorbiecki and Schneider aunties and uncles and cousins, who challenge me, goad me, and laughingly love me, and the whole Mataya extended family—*I'm so grateful to be part of such tribes!*

And most of all, to Nadja, Mirjana, and Dmitri, who have lived with this book and all that came with it for most of their lives, *which has not been easy*, and for David, who survived this with me. I am sorry for all the times I was not my best self through this long process, and I thank you for forgiving me. You are each my beloved, light-giving stars. I love you so, so, so, so, so, ... so much, to the moon and back and all around the universe.

The Canticle of the Creatures
BY ST. FRANCIS OF ASSISI

Most high, all powerful, all good Lord!

All praise is yours, all glory, all honor, and all blessing. To you, alone, Most High, do they belong. No mortal lips are worthy to pronounce your name.

Be praised, my Lord, through all your creatures, especially through my lord Brother Sun, who brings the day; and You give light through him. And he is beautiful and radiant in all his splendor! Of You, Most High, he bears likeness.

Be praised, my Lord, through Sister Moon and the stars; in the heavens You have made them, precious and beautiful.

Be praised, my Lord, through Brothers Wind and Air, and clouds and storms, and every kind of weather, through which you give your creatures sustenance.

Be praised, My Lord, through Sister Water; who is very useful, and humble, and precious, and pure.

Be praised, my Lord, through Brother Fire, through whom you brighten the night. He is beautiful and cheerful, and powerful and strong.

Be praised, my Lord, through our sister Mother Earth, who sustains us and governs us, and who produces various fruits with colored flowers and herbs.

Be praised, my Lord, through those who forgive for love of you; through those who endure sickness and trials. Happy are those who endure in peace, for by You, Most High, they will be crowned.

Be praised, my Lord, through our Sister Bodily Death, from whose embrace no living person can escape.

Woe to those who die in mortal sin! Happy those she finds doing Your most holy will. The second death can do no harm to them.

Praise and bless my Lord, and give Him thanks, and serve Him with great humility.